THE ART OF

GROWING

Premium

CANNABIS

R. K. BERNHARDT

The Art of Growing Premium Cannabis

By R. K. Bernhardt

Published by: Fertile Pages, P.O. Box 1577, Forestville, CA 95436-1577 U.S.A. www.fertilepages.com

ISBN: 978-0-692-93858-4 (pbk.)

fertile pages ™
THE ART & SCIENCE OF EPIC CULTIVATION

fertilepages.com

Contents

Target Reader

The trend to expanded legalization creates many opportunities in the cannabis industry, including cultivation. For those exploring farming possibilities all the way to seasoned professionals, the information in this book offers concepts and techniques necessary to grow crops that are competitive in the changing marketplace.

The skill level of every cannabis grower is different and so is the experience level. Just as varied are the ways to cultivate marijuana and where. There is the one common denominator of all successful marijuana farmers; however, and that is a passion to grow the best plants possible; each cycle and every season.

That endless desire of a premium cannabis grower to produce an outstanding crop is the basis for this book. The information will be a review for some readers, maybe in a new light; for others, it will be a start in the right direction or even a change in thinking to horticultural success. For all readers, there will be knowledge that can help on the path to achievement of farming business goals. Despite the scope of cultivation or the background of the farmer, this book is for every marijuana grower who wants some wisdom that will lead to a bountiful harvest of better cannabis, every time you grow it.

Forward
Knowledge for a Better Crop

Information in any field of agriculture is power and as technology moves forward, it is truly the best guide in the pursuit of excellence. In cannabis cultivation, it determines success, or failure.

Growing cannabis is rapidly changing by moving from a place of secret gardens and ordinary strains to commercial horticulture of super hybrids identified for specific properties and benefits. In a short amount of time, we have moved from marijuana seeds of unknown origin to clones with bar codes.

What is coming next is even more dramatic. The next era in cannabis will bring untold growing opportunities from small projects of private farmers to commercial ventures with significant crop production. Whether it is a large-scale operation or simply growing for personal use, the principles of sound agriculture are the same. For those seeking a profit in a crowded field of cannabis production, success will come to growers who utilize current, credible information to cultivate plants that yield the best crop, both in weight, and quality.

This book provides useful knowledge about cannabis cultivation so that you can successfully grow a top-shelf product of high value where it is legal; for either medical or recreational use, as defined by applicable laws specific to you and your location. Scientific principles of farming presented in a creative perspective combine to inspire others to pursue more information, education, research, and further development of the most important plant of our lifetimes; an herb called marijuana.

Premium Defined

Premium, despite the subject, is an ideal word to describe a sense of the finest quality and evokes a designation that is superior and first-rate. Premium implies a standard that is exceptional and the best; all excellent synonyms for how the adjective applies in the context and the crop goals of this book.

There are several factors that determine the value of cannabis flowers (buds) and each is influential in how well the crop will sell and for how much. When the nose, stickiness, color, taste, structure, weight, and chemical composition all combine in nature's perfection for the desirability of the consumer, you have a premium crop. For a grower, the materials, skills, precise methods, time, money, and work that it took to produce the premium grade marijuana determines profitability.

Some make the mistake of focusing only on one factor like weight, or appearance, or aroma, alone. Others grow only for extraction and have little concern for efforts required to grow top-shelf buds that afford numerous options for distribution and consumption. While there are many aspects in marijuana production and room for all types of growers, the fact remains that premium crops come from knowledge, attention to protocols and aiming for goals achievable in many cultivation projects. Add your own attention to details with a creative approach of always striving for perfection and the process becomes, the art of growing premium cannabis.

Introduction
A Serious Grower's Philosophy

When you first start in marijuana farming, it is an exciting time. To begin, there are choices everywhere you turn. Aside from the various culture methods available, there is a huge selection of cultivation equipment and growing supplies specific to cannabis farming. The first time you visit a hydroponics store where marijuana farmers shop, a certain amount of fervor takes over when you discover the almost endless possibilities in growing pot. You begin thinking about all the strains you might cultivate and all the money you might make. You see what appears to be smart growers throwing down their wallets for stuff you think you must need also, and the adrenaline rush continues.

For new farmers, it is also easy to believe in the myths and crazy theories about how to grow the wondrous plants along with the array of pricey products and chemical formulations promising gigantic, mega-sized buds, or plants that produce five pounds of finished product. Truth is, some of the expensive products are awesome and really work while the jury is still out on others. Adding to the startup excitement, distractions and confusion are enthusiastic opinions, and claims of success you will hear from every person who ever grew a marijuana plant in their life.

For many growers, starting a cultivation project for the first time is based on a mixed bag of information from many sources, some good and others not so great. It seems that many people claim to be experts, but it takes only about one or two poor crops to learn that cannabis will grow in almost the worst conditions imaginable and while expensive supplements and latest equipment may produce big plants from rapid growth, the quality of the finished flowers usually leaves a lot to be desired if other cultivation elements are deficient. Worse, with bad information or relying on false claims, disease, or pests conquer the effort while bad cultivation methods or practices end up wasting time and costing money.

So, the very first step before starting to grow is to come to the realization that cultivating marijuana is easy, growing premium cannabis is not. To grow top-shelf bud successfully, you should acknowledge that marijuana cultivation is a science of specialized agriculture that requires some education before you start; the more you know the better you will do. Then to really be an expert, you must continue to learn and share your wisdom with others until the day you stop growing weed. To be a truly awesome grower with consistently premium crops season after season, it also takes a certain mindset that is common among today's leading farmers, big or small.

The most incredible part of growing marijuana goes beyond the financial rewards. Serious growers will tell you that it is the process that provides the most gratification and mastering the art of cultivation produces the best results. Like artisans honing their skill set, growers of premium cannabis practice working with nature. As they perfect their craft, season after season, the leading farmers of cannabis are those best defined as artists in agriculture; for cultivation of premium grade marijuana is both a science and an art. *(Continued next page)*

Think of growing premium quality marijuana this way: like good bread from a favorite bakery, the ingredients alone do not make the quality superior and desirable, it goes to the recipe, the preparation, baking time, and the presentation of the baker who gets the final credit.

The greatness of any artisan product, like good cannabis, is the result of combining the components, the technique, talent, and most of all, a passion for the result. So too, the information in this book is merely an ingredient list with recipes, carefully seasoned with ideas, and inspiration to guide you on a path to abundance. It is up to you to mix in the unique and artful ways you alone can passionately grow the best marijuana, ever.

"Agriculture is the foundation of manufactures, since the productions of nature are the materials of art."

-Edward Gibbon

Chapter 1

Explore & Plan: Develop a Vision For Success

Growing cannabis can be one of the most fascinating experiences you will discover in horticulture. For hobbyists or private growers cultivating for their own use, it is rewarding on many levels, and a good way to save on the product. For more serious or commercial growers, it is an avocation that requires a professional approach for any degree of success. Despite the size of your garden or your project, it is worth spending some time checking things out, before starting or expending any money on a cultivation endeavor.

Exploring all your options should include reading and digging deeper in the areas that interest you the most, whether it be propagation, growing, packaging, sales, etc., just to learn all the possibilities. Then visit your local hydroponic store or a nursery that specializes in supplies for cannabis cultivation (they are out there, just look and ask around and you will find them). Examine equipment, supplies and prices. The Internet also provides a wide range of information, from strain statistics, to photos, to shopping, you name it. If you plan to grow medical marijuana, your local dispensaries offer a wealth of assistance from books to advice, including recommendations of local vendors to help when beginning.

Planning plus execution equals success. Knowing what to expect about requirements and expenses before starting will make your farming ventures easier and more profitable. Investing your time before your money is the best advice in this book.

1 | a.   Know What You Are Getting Into
Farming Challenges to Consider

In conceptual terms, consider your marijuana garden as a scientist would think of growing in a dome in any environment, on any planet. The better you create conditions that are ideal to your farm, the healthier the cannabis plants will be and the better the yield at harvest time. However, despite the opportunities afforded a premium marijuana farmer, cultivation can be as risky as living in a glass house. Good planning and best practices help to minimize the challenges.

Working with any living entity presents a varied amount of demands that change constantly. Managing those aspects in crop cultivation while working with the unpredictability of nature is what agriculture is all about. Beyond the rigors of basic agriculture, growing cannabis also has a unique set of factors unlike other consumable crops, in that it is highly regulated. Laws control the legality of cultivation, how and where to cultivate and further regulates the distribution of the finished product.

Highly desirable with cash value, cannabis is also subject to theft at any stage of development while growing and during processing, distribution, and sale. So, while a farmer of corn may not need to think about providing tight security to his row crop, a marijuana farmer must think about it always.

Maybe the most under-estimated part of growing marijuana is the labor required throughout the process. To grow more than a few marijuana plants successfully is usually a full-time job during a crop cycle; today's marijuana hybrids require attention from a seedling to maturity and from the harvest to final sale. Clearly, farming cannabis must be a commitment of time and effort that takes priority over everything else if you want to be grower of top-shelf bud.

FIXED COSTS & VARIABLE EXPENSES

After the initial setup expenses, the importance of identifying growing expenses before you plant anything is critical to your success, and funding a reserve for unexpected costs is an excellent financial strategy. If your focus is on your plants in culture and not on money worries, everything goes much smoother on a path to premium weed, even if you are growing for personal use.

Fixed costs are expenses that you incur despite crop production and of interest to commercial ventures. Examples of fixed costs include property taxes, mortgage payments, lease payments, rent and salaries. You should start with enough funds to make these payments for one grow cycle, especially if you are just starting. Hobby growers also need to think about these expenses if you are not growing at home as they will reflect in the cost of producing your premium product.

Variable costs are expenses that increase or decrease with crop production phases. Examples of variable expenses include labor, utilities (water, electricity, gas) seed/clones, fertilizers, insecticides, substrates and more. Determining the estimated totals for variable expenses is specific to the type of cultivation method and units (plants) in culture, but the funds necessary to have available should also be enough money to complete at least one vegetative and bloom cycle. As in many forms of agriculture, do your best to avoid the use of credit for any farming expense; the risk of crop failure is not worth interest-laden debt.

WEATHER & CLIMATE CONTROL

There is specific temperature, humidity, light, and air requirements for growing marijuana and particularly for hybrid strains. Creating and maintaining an ideal environment for the plants is not always easy. When growing outdoors, adverse conditions can be difficult to remedy, while indoor climate ranges can be costly to maintain at optimum levels. The climate requirements for growing premium grade marijuana are stricter and a consideration when becoming a grower.

PESTS & DISEASES

There are numerous ways to prevent and control pests and diseases, safely, legally, and in consideration of the environment. A simple solution is to only use products approved for use on food crops and listed as safe or approved for use on cannabis; there are many available. As testing becomes the norm and as the extraction part of the industry booms, you can avoid many contamination issues resulting in crop loss just by growing in methodical procedures of "food-safe" standards.

Even with strict protocols, marijuana plants are the subject of great interest regarding pesticide use by The US Environmental Protection Agency (EPA), US, State, County and Regional Agricultural agencies, Water Boards, and other licensing agencies specific to cannabis. Compliance that includes permits and paperwork will likely be part of the process for more than one entity because of the scrutiny. Many jurisdictions control the application of pesticides by a permit process, with restrictions that may include storage standards, inspections, and training. Consider the time and expense for these and other regulatory procedures as you examine cannabis cultivation requirements and your own personal resources of time and money.

COST OF LABOR

Specialized plants like marijuana do not rely on mechanization like other farm crops and most cultivation methods require manual labor to plant, tend, and harvest. Each task associated with growing premium cannabis requires a bit more attention to detail than the average crop and so, requires even more hands-on labor.

Labor costs can vary between salaried work and hourly compensation valued by regional rates for specific work. For example, "The going rate" for bud trimmers might swing widely from one area of the country to another, or from one part of a state to another. Check with local or regional organized growers or supply vendors to find workers and to assist in determining the expected labor cost your farm might incur for the services you need.

FOLLOWING THE LAW

Compliance with all laws and regulations pertaining to cannabis is the best policy for any grower. The legalities may require you to spend money to become or remain legal; an expense many farmers do not plan for. The industry will continue to flourish providing it is compliant and professional in all regards, and it is the serious farmer who should lead the way. Planning to grow credibly with long-term success means investigating what the laws are and following them to the letter. Enough said.

While farming premium marijuana is not easy, can be costly, stressful, risky, and laborious, any successful grower will tell you, it is some of the most satisfying work you will ever do. Knowing ahead of time what you might face and planning accordingly will make it all the sweeter.

Essential Resources, Facilities, and Supplies
The Basics for Growing Cannabis 1 | b.

To grow healthy, and robust flowers require indoor farmers to create an ideal environment for cultivation and provide everything a healthy plant needs through harvest. With good lighting and air climate control, plants will require grooming, staking, irrigation and nutrient application.

For greenhouse or outdoor growers, the plants still require optimal provisions for production of great crops with an entirely different set of challenges than indoor gardening. In all types of cultivation, indoors or out, the plants require optimization of physical conditions where ever possible to do well and produce a quality crop.

It is unlikely that any book can provide you with specific advice to your individual cultivation project, but a basic picture of requirements can serve as a model for your planning and customization. Growing premium grade marijuana is not easy. It takes an investment of time and money like any form of agriculture, but it also requires planning not found in other types of crop production. For that, the expenditures required to grow top quality bud should be your first concern while giving careful thought to becoming a marijuana farmer.

The economics of growing premium cannabis is important because you do not want to waste money growing it if it is less costly to purchase it from a legitimate source. As legalization increases, so does production, requiring markets to adjust initially with increased supply and lower prices. Until markets stabilize, it is foolhardy to grow cannabis for any financial gain if the market has product that is selling for less than your crop cost to grow. Growing for personal use may not need any financial consideration, but cultivating cannabis for legal sale does.

To the point, planned profit and loss is one thing; actual essential point numbers are another. For that reason, keep track of your expenditures related to actual cultivation of a crop (not setup expenses) and measure the weight of the finished product to determine the approximate cost of production. Compare it to local market rates for a similar quality and decide the benefit to grow it yourself, for personal or commercial use. The self-examination year after year is a smart assessment of your true success.

What you need to grow premium cannabis mirrors what you need to grow most forms of the plant, despite the end quality. While your operating costs will vary with each crop and by cultivation methods, your startup expenses are essentially the same; you need certain items to grow marijuana properly. Once you have the foundational requirements, you can enhance your equipment and supply lists to meet your premium crop goals. There are five key areas of requirements you should examine as you consider marijuana cultivation, indoors, in a greenhouse, or outside in the sun; accommodation and improvement of each category should always remain a goal.

BOTANICAL

Marijuana plants require air, water, warmth, light, carbon dioxide, nutrients, container, and substrate to support a growing plant.

CULTIVATION

A source for seeds or plants, space for growth, electricity to operate equipment, lighting, climate control equipment, irrigation system, air, and water monitors/meters, growing containers, and supports, tools for pruning and propagation, supply for pest control, hygiene, and sanitation is a list of cannabis farming needs.

PRIVACY

In the simplest terms, the fewer people who know about what you are growing, the better. Cultivation areas should be out of sight and not within plain view of the public. It is a good idea to refrain from showing your garden or talking about what a fantastic grower you are to anyone who does not need to know about it, even trusted friends, and relatives.

TIME

Growing marijuana requires much labor for monitoring, irrigating, fertilizing, grooming, pruning, spraying, staking, etc. Harvesting also takes labor for picking, trimming and storing. The more time you allocate to cultivation and harvest, the better your result, but do not consider it a part-time exercise if you want quality results. Many top growers spend sixteen hours or more per day tending to their crops, aside from hired assistance.

FINANCIAL

To do it properly and achieve premium quality results, the cultivation of cannabis takes money; from setup to continued purchase of growing supplies like soil and nutrients. Do not think that you will recover your expenses with the first crop either. Seasoned professionals know that to do well a farmer must be consistent crop after crop and rarely does anyone strike it rich on the first try.

Before you start, remodel, or expand a cultivation project, speak with other growers. Ponder if growing marijuana will meet your expectations with the various commitments you must make. If you have thought it through, have the dedication and passion to go forward, then learn as much as you can and go for it; you will never be sorry you did.

Farming Necessities
tools & supplies for efficiency

There are hundreds of worthy things that a marijuana farmer could purchase for cultivating the wonder plant. Then there are things that you really need to grow premium cannabis. This is a list of those items; tools, equipment and supplies that you should have for the job of growing an elite crop. In review, also think about an organized system of storage for your cultivation tools of the trade.

• Basic gardening tools; too numerous to list, but always buy the best you can afford and take care of them. A good quality shovel and an ergonomic trowel are must-haves.

• pH Pen; single most important instrument you will own. Choose one that has a high degree of accuracy and measures water/solution temperature also.

• Needle point shears; a tool that needs sharp blades so it may pay to replace at the start of each season. You will use this tool frequently from the start to harvest trimming. Clean with rubbing alcohol and then wipe with hemp oil to keep it clean.

• Thermometers and hygrometers; buy reliable instruments to measure temperature and humidity (price does not equate quality or accuracy) and place at canopy height near plants, inside or outside. Although you may have little control over the climate of an outside garden, you should be aware of the conditions.

• Large watering can; a handy and most useful item for quick dispersal of water or nutrient solution to correct deficiencies. You will use it often.

• Mixing and measuring containers; plastic is readily available and does not break in the workplace, but glass or metal are better for the environment. Purchase several sizes to sample, mix fertilizers and measure chemicals for use in nutrient solutions and pesticide sprays. Five (5) gallon white paint buckets are most useful; purchase with lids for a variety of uses.

• Gloves by the box of one hundred; powder-free nitrile gloves (disposable) are available in different weights and specific sizes for comfort and dexterity. You will use a lot of them, so stock up on a type that fits well and you like.

• Pest sticky traps in blue and yellow; not only a good way to catch pests like fungus gnats, thrips, and whiteflies, but also an excellent method of monitoring activity. Place near plants on stakes or suspended, but make sure not to let the sticky portion touch plant foliage, a possibility from air movement.

• Good quality tank sprayer; two and one-half (2.5) gallon is a good size for portability, but larger types save time and labor. Keep equipment clean and it will last for more than one season.

• Notebook and plant labels; sounds basic but awesome tools for tracking and monitoring your plants from seed to final sale.

• A microscope; last on this list, you really should have a microscope if you are a serious grower. You will use it to identify pests, nutrient deficiencies, and examine resin glands for maturity.

Plastics and Pollution

There are serious problems facing our environment and specifically the oceans. Since plastic takes generations to break down, scientists estimate that by 2050, there will be more plastic debris in the ocean than fish. For anyone, that is a profound proposition.

Waste reduction is a large part of the solution to this global issue, but since only about 9% of plastic gets recycled, there are other effective ways to turn this pollution nightmare around. As a consumer in the marketplace, your purchasing decisions drive how manufacturers respond, so the first line of defense begins with you, as a small farmer or large commercial grower.

Make certain to recycle any plastic waste, but when you can, buy products made and packaged without all the plastic stuff we are accustomed to. Use glass and metal for measuring, as one small example that can make a substantial difference, especially if more have the mindset to consider everything we do has an influence on the world in some way.

A leader in this effort is the Monterey Bay Aquarium in California. For more information, see the Science & Conservation section on their website:
http://www.montereybayaquarium.org

Finding Credible and Reliable Sources
Value Good Information, Suppliers, and Customers

Some of the most knowledgeable people in the industry work at dispensaries. Their cannabis wisdom coupled with a retail experience puts them high on the list of important contacts that marijuana farmers need to make. The best way to get to know them is to support their enterprises with purchases and referrals. The cannabis business is all about networking and starting locally is a good first step.

The world of cannabis cultivation is an interesting one, to say the least. Like many aspects of life, there are good folks in the marijuana business and there are some outlaws. Aligning yourself with the brighter side of the coin will help ensure your success and more important, your safety.

There are several venues that will be supportive and helpful to your cultivation efforts. Among them are medical marijuana dispensaries, co-Ops, trade groups, and associations, nurseries, and hydroponic stores. Add to the array of information and equipment resources available to you locally is the Internet. Although you cannot believe everything you hear or read on the topic of cannabis cultivation from any single source, a consensus of acceptable and valid information will begin to emerge the longer and deeper that you investigate.

Many places that offer sound advice to you will also be aware of interested parties in your finished product, or can steer you to legitimate customers who may be. Like all things cannabis-related, be discreet, and follow the law.

With marijuana laws changing nationally and internationally, businesses to serve the emerging markets are popping up everywhere. Many are great resources; others are clueless. Before you select where you will do business and what recommendations you will value, shop the marketplace. That means visit cannabis related stores and businesses within a 25-mile radius of your location. Compare prices, service, and your overall impression. Feel free to ask with questions and expect honest, open answers in your evaluation process. Marijuana trade shows are also an awesome resource.

If you are a medical patient or reside where recreational use is lawful, visit as many dispensaries and retailers as you can, and buy products! By supporting a cannabis enterprise, you are investing in your chosen industry and your own future while building healthy business relationships. Next, look for Internet retailers and websites of manufacturers who make products you intend to purchase. The wealth of information to assist you in making informed buying decisions is deep and wide.

PURCHASING PITFALLS

Finding great places to spend your money is not always easy, but developing a working relationship with vendors you can trust is a wise approach. Fair prices (the cheapest is not always the best), dependable service (for knowledgeable advice), availability (location, and hours of operation) and most of all, trustworthy in all regards are traits you must rely on for customer satisfaction.

Even with places that you trust, there may be issues to watch out for:

•Soiled or damaged products returned to the retailer after use for resale.

•Nutrients returned with broken seals may contain diluted contents.

•Products might have reached or are past their effective shelf life.

•Seeds may be old or stored incorrectly for viability.

•Products with living material past usefulness from old age or improper storage.

•Clones with pest or disease exposure (mites a common culprit).

•Mislabeled seeds or nursery stock (clones and plants).

•Internet or Special orders with huge shipping charges (heavy weights and bulky sizes equate to additional fees).

•Service providers looking to "cash in" on new proprietors or naive farmers.

SELLING THE FRUITS OF YOUR LABOR

Assuming you have a product that you can be proud of, it is most useful to have it tested by a lab that specializes in cannabis analysis, before you offer it for sale; for private or commercial use. A retailer appreciates a chemical test or bioassay of a sample of your crop with information (chiefly THC and CBD content) that will help immensely in selling your premium weed, while assisting in attaining top-dollar for it is true worth on the market. A "clean" test resulting from no residue from chemicals, pesticides, fungicides, or mold is becoming a common requirement for cannabis products including extractions; so, testing before innocently trying to sell a bad product is smart business. Rates for testing can vary from one laboratory to another, so compare services.

Your next step is to plan your exposure to the marketplace. Locate dispensaries in your area and authority and investigate their farmer requirements for standards of quality and sampling. Then go visit them! Talk to your potential customers and find out what they want, not what you have to offer. Although most will sell or package your product in their containers, there are a few steps you can take to help you sell your premium crop, even in bulk transactions.

Presentation is everything, and anyone who tells you it is unimportant is not long for the cannabis trade. Even when submitting samples for buyer's consideration, get creative and professional with your product container. For distribution to a consumer market, packaging and labeling requirements varies by authority, but the best suggestion is to go overboard; give the consumer more than they expect in reliable, tasteful, and informative product packages. In a competitive marketplace, it is the edge you need to get your goods sold. As time progresses, branding and your identity will become vital to your farming success and the sooner you start the better.

Labeling requirements also varies by locale, so make certain to include information like warnings for safe use, keep away from children and pets, etc. Include strain information, cultivation method of the product, location, and the weight. Bar codes are useful for inventory management and point of sale systems even where they not required.

Prior to sending your goods to market, do some price comparison to determine what it is worth. You may think it is the best stuff around, but ultimately the end user determines the desirability and price. Use the same strains if possible when analyzing and comparing value, or those with similar quality and THC content. Get realistic about what price to expect, if your finished crop stands a chance of sale against other growers in a very competitive marketplace.

STRAIN SELECTION

Deciding the type of named cultivar to grow is an important consideration for a premium crop. Hundreds of named varieties are available so your choice must consider several factors, including aptness to your culture method. Trying to grow a hybrid that is best suited for indoor cultivation, out in the elements under the sun, will likely lead to a less than desirable harvest; both in weight and quality.

Start the selection process by researching various Internet sites that index current or known strains and create a list of those that interest you (if you are not into the bud, your passion for perfection will not be at 100%) and look for those that will grow well in your chosen garden space. That might be indoor, in a greenhouse or outside, commonly called sun-grown. Fortunately, most reliable strain indices use the location or culture method as a primary reference, often indicated with a sun icon for outdoor, a light bulb icon for indoor or a house symbol for greenhouse growing. Some even categorize or filter strains by the attributes they provide.

Your next step is to decide the marketability for your chosen strain if you plan to grow for commercial distribution. You do not need to be exclusive, since unknown, or uncommon strains have a slow start before market forces allow consumer familiarity of the type, but you also do not want to grow what everybody is growing either. Ideally, you want a strain that you enjoy, has some ease in growing for your method of cultivation and has the potential to be a good seller. If you are growing strictly for extraction of the buds, your careful thought about strain selection is no less important than a grower of flowers for combustible consumption. Chemical compositions of finished flowers should be your principle concern despite the final or intended uses.

Individual hybrids have characteristics that make them desirable for a medical patient and pleasurable for a recreational user. Indica dominant strains tend to be more sedative than sativa strains that tend to be more uplifting, while cross-bred strains might exhibit uniqueness with a little bit of both parents in the hybrid. Strain references often list in detail the characteristics according to the genetics, but some breeders either are uncertain of them, choose to keep them private, or else the information is unknown.

Also, be aware that there is some deception in the marketplace by unscrupulous breeders that label their clone stock incorrectly; not a lot, but it happens. Please label correctly if you supply other growers who may have a livelihood that depends on an accurate strain selection at the start of their grow.

Find Your Ideal Cultivar:
For strain exploration, selection or sourcing, check out Leafly.com

A chief reason businesses fail is that they are under-capitalized and the same holds true for cannabis farmers who grow premium marijuana for a living. Operating budgets and long-range financial planning help minimize the pitfalls of not having enough money to operate, expand, and prosper. Put pencil to paper and execute your financial farming plan to a bountiful future.

Growing marijuana as a hobby or past time, for personal use, or for sale where it is legal is an activity that requires expenditures. For vocational growers, plan, and account for these expenses just as a farmer who grows wheat or corn manages his farming budget. Knowing what your crop costs to produce is the best way to determine if your effort is worth it from a financial perspective; it is also a useful tool to plan pricing goals and look for ways to control costs for future cultivation projects or seasons.

An expense and budget planning process can go as deep as needed broken down into three major categories:

•Facilities or Land Expenses: What does your site, greenhouse, or indoor building cost to purchase, rent or lease?

•Equipment Expenses: What is the total cost of durable goods or supplies required for cultivation and useful for successive growing over a one to five-year period?

•Cost of Goods Sold Expenses: What are the operational costs required to produce one crop or over an accounting time, like a year, or season?

BASIC START-UP COSTS

It will be challenging to produce premium grade marijuana unless you have the resources to complete one entire grow cycle properly, so as a practical rule of thumb, know what you will need beforehand, acquire those assets, then begin to grow. It is essential to be attentive to your plants in cultivation and not worry about funding, especially in the middle of cultivation. Starting prematurely without a plan and the money to execute that plan effectively will lead to certain failure of any business proposition and that includes growing marijuana, especially premium grade.

There are so many growing options and attractive items available for cultivating premium cannabis that it is very easy to break the bank and spend money foolishly, especially when starting. The task of knowing where to start any planning process can also be daunting without a little guidance, so begin with these steps to obtain some key numbers:

1. Select your cultivation method and location.

2. Decide how many plants you will cultivate in the allocated space for your method.

3. Multiply the number of plants by the starting cost per plant. (Plant/seed cost) + (the container cost) + (the grow media or soil cost).

4. Determine the nutrient (fertilizers, additives, etc.) required per plant for all stages of growth.

5. Multiply the nutrient costs for an entire grow cycle by the total number of plants.

6. Select your primary pest control method and list the material costs for one grow cycle, plus application equipment.

Except for labor requirements, calculating these expenditures, based on a cost-per-plant formula assists in finding most of the material expenses to produce one crop.

UTILITY EXPENSES

Often under-estimated or even overlooked by new farmers, the cost of electricity can be a significant portion of your operational expenses. Utilities include the costs associated with climate control (heating, cooling, humidity, air movement) water, and nutrient delivery, pumps, etc. This is of concern to indoor growers who must maintain an ideal growing environment by artificial means, or when using greenhouses to their full potential.

When calculating anticipated costs, it is important to know your estimated power consumption needs for a one hour period, measured in watts if your source is electricity. That means adding up every item down to the last bulb or water pump that you will use in cultivation. Listed on all appliances and fixtures are the power consumption, or obtain it from the manufacturer.

There is roughly an average of twelve to sixteen (12-16) weeks in a total growing period for most cannabis strains, although the range can vary widely. If 12 that equates to 84 days; or 2,016 hours of total power use. Multiply your watts per hour needs by 2,016, then compute your local cost per kilowatt hour to obtain an estimated cost of electricity. (Current National Average is about $0.12/kilowatt-hour; your utility company can provide your exact cost and assist in planning for energy cost efficiency.) Also, commercial accounts usually receive reduced rates over residential customers.

(Wattage x hours used) / (1,000 x price per kilowatt-hour) = cost of electricity

Example:
(total watts per hour x 2,016) / (1,000 x 0.12) = expected cost of electricity

MISCELLANEOUS SUPPLIES

With all the merchandise available to assist in growing premium weed, there are so many variables; some things you need and some that you just want. When calculating your starting costs, buy the best you can afford and purchase in bulk when possible. As your operation enlarges, your buying and saving ability increases as well, with discounts from larger purchases and negotiated terms.

RESERVE FUND PLANNING

Despite how meticulous your financial plan may be, there are always unexpected expenses that will pop up during your cultivation project. So, set aside at least 10% of your expected expenses for one grow-cycle or crop to meet whatever challenges may arise in growing your premium cannabis.

Profit and Loss Realities

There is no easy road to riches in any part of the cannabis industry; least of all in cultivation. Any successful farmer will tell you that when you are working with living material, there is risk of failure. Sometimes, all it takes is one bad crop to wipe you out financially. To succeed in growing cannabis, you need determination coupled with optimism. Most of all, you need to possess a willingness to work hard and smart and that starts with a plan and a periodic review of the financial part of your cultivation business.

Plant Strain Protection

Cannabis farmers have many responsibilities just in operational oversight. But beyond our own farms and gardens, the future success of this industry relies on the bigger picture. Looking forward, lead growers need to encourage a culture of community; one in which science based farming sits at the head of the table. From seed, to production, to processing and then to distribution, integrity of the product needs to rule the process.

Every grower should consider themselves a cannabis scientist, with a role of stewardship over the many named strains and their origins. Essentially, it is all about the name, making basic protocols of correct labeling and identification imperative, for so many reasons. Research and development by breeders rely on this accuracy, but more important is the need for exacting protection of existing cultivars. It is what science requires and what serious growers with much money on the line depend on. They need to know that what they are growing matches what the label says.

The Agricultural Marketing Service of the United States Department of Agriculture oversees The Plant Variety Protection Office (PVPO) that provides intellectual property protection to breeders of new varieties of seeds and tubers. While cannabis is not currently a protected plant under The Plant Variety Protection Act (PVPA), the program encourages development of new and improved varieties better suited to climate change as well as better pest and disease resistance. The PVPA protects breeder's stock in three ways; Plant Variety Protection for 20 years (25 for vines and trees), Plant Patents for asexually propagated plants, and Utility Patents for plants that show utility. This legal protection is not available to cannabis strains, but it serves as a model for future use. It also helps explain why it is important to any farmer now and in future generations and offers guidelines for cannabis growers to get it right.

For more advanced information on the protection of plant varieties, dig as deep as you can; the reasons are very significant for the future of agriculture, despite the crop. Maybe one day, cannabis will also receive these legal protections. Until then, it is up to professional breeders and growers to develop and follow our own rigorous standards.

As regional approval of cannabis cultivation occurs, more strains become available on the market, no matter where it happens. Part of it is marketing craziness, but most of the new strains are the result of careful and methodical science; the kind that good breeders rely on and practice. It is also why it is wise to purchase your cultivation stock, seeds, or clones, from these reputable sources. At the end of the day; however, it is the responsibility of everyone in the chain from the seed to sale, to be ethical, honest, and accurate in the labeling and protection of cannabis strains; after all, it is our future.

Chapter 2

Cannabis Agriculture: Knowledge for Ultimate Growing

Sciences of interest to agriculture are also applicable in the cultivation of cannabis. The farming or cultivation of marijuana sometimes called *Cannaculture*, relies on principles of producing crops just as a farm growing food for our table does in an endless pursuit of bigger, better yields. While you do not need a degree in crop science to do well growing marijuana, smart cannabis growers are also on a never-ending journey in their pursuit of knowledge about how to grow an improved bud.

Just like any successful farmer of a food crop, you should also constantly seek current wisdom about growing marijuana. You must; the strains cultivated today require more precise culture techniques and the materials to do so are not cheap. Like any commodity, cannabis farmers sell, trade, and distribute in markets for profit. Within this competitive arena that gets more populated continually, only the best quality products, grown the most efficiently will succeed.

This Chapter introduces you to the science of soil with a focus on how it applies to growing healthy and productive marijuana plants. If heavy science is not your thing, use this portion of the book to gain an oversight into the basics of the plant, some agronomy, and as a source for future reference when you are ready to invent your own super growing medium.

2 | a. Basic Marijuana Botany
A Genus of Flowering Herbs

Cannabis sativa

From origins of open pollinated land races to highly developed strains with traits selected for breeding and cultivating, cannabis has come a long way. Gone are the days of planting bag seeds and hoping for the best. Today's cannabis farmers are precise and methodical in the pursuit of the perfect flower; defined by the market.

Referencing the taxonomy and understanding the anatomy of a marijuana plant are pursuits worthy of any cannabis professional and particularly a farmer. Catering to the optimum requirements of all the parts of the plant will have beneficial results for any grower, despite the cultivation method.

Native to the Himalayas and indigenous to Asia and India, cannabis is an annual, dioecious plant in the family *Cannabaceae* that include three species; indica, sativa, and ruderalis. Although grown for centuries for fiber, essential hemp oils, as well as for medical and recreational use, the subject matter of this book focuses on the plants selectively bred to produce high levels of THC or CBD; primarily from indica and sativa species.

The leaves of cannabis are very recognizable with a venation pattern that is distinct and typical. Variation in size shape of leaflets is common in all species, including those plants bred for low THC content and cultivated for fiber, called hemp.

Root systems on cannabis plants vary by species, strain, and method of propagation. Usually, plants started from a seed will develop a distinctive tap root. Plants started from clones will typically develop a system of more uniform branching from the cut site. Like all aspects of nature, there are exceptions and variations to these generalized root system descriptions.

The flowers of cannabis plants are not identical and identified as imperfect. Male flowers (staminate) [see facing page illustration items 1, 3] and female flowers (pistillate) [illustration items 2, 4] occur on separate plants. While it is rare, some plants will bear both male and female flowers, especially in new strains where mutations can appear from hybridization. This abnormality, often identified incorrectly as hermaphrodite since true hermaphrodites, where male and female portions appear on the same flower, rarely occur in cannabis.

Trichomes containing cannabinoids and terpenoids are glandular structures on the plant. Those that grow on the calyx [illustration item 4] are of most importance to the grower, appearing in a more concentrated and usable form than those that appear proximate to the blooms, like adjoining foliage or stems.

Natural pollination of known strains is anemophilous, or by the wind. In the breeding laboratory, breeders use mechanical means of pollen transfer for seed [illustration item 5] fertilization and reproduction. Breeders select traits from more than one strain and cross-breed them to achieve a strain that is desirable for specific reasons like taste, smell, effect, etc. When strains are cross bred, it is often to achieve hybrid vigor, where the offspring grow better than the parent stock. Breeders also look for genetics that are fresh to avoid excessive inbreeding of the same seed stock.

Premium grade marijuana is achievable from plants grown from a seed, or from a common and popular form of propagation called cloning. Clones are simply cuttings from a desirable mother plant, rooted in soil, or grow medium, and grown into full maturity, retaining the characteristics of the mother plant. The advantage to growing from clones is uniformity in size and maturation of the plants, particularly useful in larger cultivation projects where breeders prefer and select specific genotypes.

During the early bloom phase of development, trichomes and pistillate hairs form on flower buds in this close-up of a female sativa species in cultivation. In simple terms, this is the crop for a cannabis farmer.

Most strains of cannabis are short-day plants that produce flowers (buds) when the night length is longer than their critical photo period. Growers of premium grade marijuana often utilize this characteristic for predictability of outside crops because of daylight hours or for controlling bloom cycles in crop production when growing inside.

The plant structure of cannabis species varies by strain or phenotype and is a factor in selection for space considerations. Some types grow like a Christmas tree; others like a squatty bush. There are some that reach heights of twelve (12) or more feet and depending on the cultivation method, some may only grow to a height of three (3) feet. The outer edge of the foliage from the trunk is the drip line. This distance varies by strain as well, often directed by early training of young plants by a grower. The term canopy refers to the top portions or outer layers of a plant or group of plants.

Trunks can vary in diameter, from one-half inch to three (3) inches or more, although larger diameters usually only appear in stock that is growing in a field or cultivated in very large fabric containers, once used only for trees. A longer growing period of the vegetative phase has an influence on trunk size also.

The distinctive shape of a marijuana leaf helps for easy identification, but the leaf part of the plant is the most sensitive to climate conditions of light, air, temperature, and humidity, and important to a farmer for observation since it shows indications of plant health and vigor. The leaf portion of the plant is also the most attractive to pests that prey on cannabis.

In the final steps of growing premium cannabis is the drying and curing of cannabis flowers, at the perfect time, in ideal conditions that produces the finished, desirable

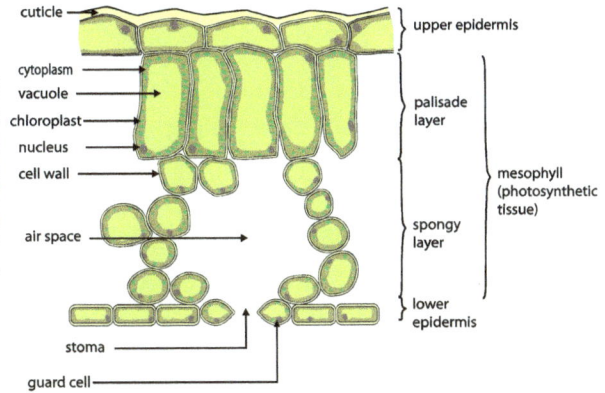

cuticle — upper epidermis

cytoplasm
vacuole
chloroplast
nucleus
cell wall

palisade layer

mesophyll (photosynthetic tissue)

air space

spongy layer

lower epidermis

stoma
guard cell

The complex structure of a leaf supports a range of vital plant functions necessary for growth including photosynthesis and transpiration. Consideration of leaf morphology will help in good cultivation techniques for better plant health that will lead to higher crop yields.

product for medical or recreational use. Small and fragile, just a little larger than a comma in this text, trichomes must ripen on the plant while growing, then harvested by removal of the whole flower buds from the plant for drying.

Maximizing bloom size and tweaking the chemical content therein are chief goals when growing premium grade marijuana. The trichomes appearing on the blooms are a measurement of cultivation success. Their perfected structure and composition are the goal of cultivation and the focus of final production because they contain the compounds that determine quality, weight, and value of the flower. In short, growing premium grade cannabis is a careful process leading to desirable flowers. To reap the real bounty of the crop, those flowers must be the object of attention until final sale.

More weight and larger blossoms usually lead to more trichomes, but also require additional consideration; during cultivation, supported as they ripen, protected from pests and diseases that attack the blooms, picked at their peak (when the trichomes are "ripe") and handled carefully through the grooming, drying, and curing steps; all in the pursuit of premium weed.

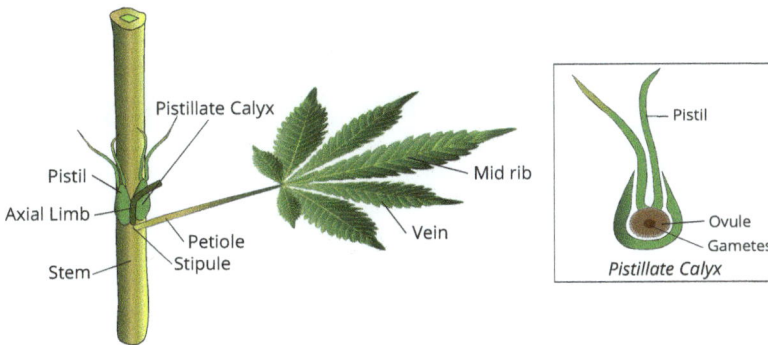

Pistillate Calyx
Pistil
Axial Limb
Stem
Petiole
Stipule
Mid rib
Vein

Pistil
Ovule
Gametes
Pistillate Calyx

The female cannabis plant is what farmers grow for flower and resin gland production depicted in this plant section. A magnified cross section of a pistillate calyx shows where a seed would form (ovule) if pollinated from the spores of a male plant.

2 | b. Use the Sciences of an Ecosystem
Soil Wisdom Is the Pro-Grower's Tool

The substrate employed to cultivate cannabis holds the key to crop quality and yield. The more you know about agronomy, the better understanding you will have in selecting a suitable substrate, amending soil for planting or designing the ideal growing medium for a bountiful harvest.

Maybe the single most important factor in cultivating top-shelf bud is the grow media provided for the plants. Despite the cultivation method, if the growing substrate (everything from ground soil to soilless mixtures used in hydroponics) is not optimal for cannabis, you will not get the top yields; either in quantity or quality. Choosing the best growing media, and then maintaining it properly for the best crop production requires knowledge of the sciences concerned with soil and agriculture.

Like most cannabis cultivators, you want to grow impressive plants with big, beefy buds, not get a degree in soil science. The information provided here in a horticultural context is to acquaint the reader with concepts helpful to grow premium cannabis. With a basic understanding of agronomy, cultivators can work with any substrates and obtain the most desirable results. Whether it is native topsoil or soilless media, your goal should create a living ecosystem that will nurture the full potential of your plants. To do so correctly, you should not only become familiar with the basic sciences pertaining to soil and plants to manage wisely, but to also develop and optimize a specific cultivation plan for your crop production.

FUNDAMENTAL AGRONOMY

Containing a mixture of minerals, organic matter, liquids, gases, and living organisms, soil is the most abundant ecosystem on earth. Although soil covers only about 10% of the planet's surface, it is a global carbon reservoir rich with geologic and climatic history. Microorganisms including fungi and bacteria populate the soil, with estimates ranging from 50,000 species to 1 million in a single gram, from locations across the globe.

Minerals from weathered rock, called the parent material, are the primary component of soil. The source of the parent materials can be the bedrock below the soil, or from other locations transported to the soil by natural forces like gravity, water, wind, or ice. Parent materials influence soil properties and the microorganisms that break down minerals and organic residues. Topography, or the shape, and geology of the landscape contributes to the composition of soils, along with the climate that affects various soil properties. Add to the mix of interaction and change is vegetation that affects soil characteristics in several ways.

Soil is a natural product of climate, relief, biological action, and the parent materials interacting over time; always in a state of development. Human changes to land from activities like agriculture add additional factors that influence soil. From a scientific perspective, detailed descriptions include texture, water content, chemistry, structure, color, porosity, consistency, reaction, and so on. A general understanding of the criteria for evaluating soil helps you select the correct growing medium and amendments.

SOIL PROFILE

The vertical slice or section from the ground surface downward to the bedrock is the soil profile. When growing cannabis outdoors, the profile of a planting area is important as it varies widely by location and has a direct effect on plant growth.

Varied size particles of the parent material distribute in layers of a soil profile, called horizons. The topsoil containing organic matter is Horizon A. The depth and properties of this horizon are of concern to the outdoor, cannabis grower who plants directly in the ground, even when using planting bags of material designed to bury in planting holes at various depths. The loss of materials, or leaching, is the movement of water and soluble matter downward to the bedrock. These materials accumulate in the subsoil termed Horizon B. The C Horizon is below that, consisting of a mix of subsoil and parent material. Horizon R is the bedrock, or parent material. Horizons and their thickness vary greatly by location. Soil scientists called pedologists study soil profiles and horizons for classification and interpretation of soil use. Some properties of horizons are clearly visible in a cross section of terrain, like color, structure, thickness, etc., and some require laboratory testing to define specific soils.

Capital letters O, A, B, C, and E, identify the Master horizons, and lowercase letters for distinguishing qualities of these horizons. Most soils have three major horizons; the surface horizon (A), the subsoil (B), and the substratum (C). Some soils have an organic horizon (O) on the surface, but this horizon often occurs buried and unidentified. The Master horizon, E, is for subsurface horizons that have a significant loss of minerals, called eluviation. Hard bedrock, which is not soil, uses the letter R.

} **O Horizon**: 2" depth*
Organic matter; loose litter, partly decayed

} **A Horizon**: 10" depth
Mineral matter and humus mix

} **E Horizon**: 30" depth
Zone of eluviation (mineral loss) and leaching

} **B Horizon**: 36" depth
Subsoil accumulation of clay and minerals

} **C Horizon**: 48" depth
Substratum of partly changed parent material

} **R Horizon**: Varied depth
Hard bedrock

*Soil depths vary widely; provided for comparison.

AGRONOMY AND OUTDOOR CULTIVATION

Any landscape can have several different soils, each with complex properties. These properties can change soil use, especially as it applies to agriculture. Some properties change with the addition of amendments, including organic material, minerals, and nutritional supplements. Other characteristics such as topsoil depth and drainage capabilities are largely unalterable for growing cannabis and are factors to consider when choosing an outdoor site.

Criteria for selecting an in-ground planting site should start with a location that has maximum sunlight exposure with ample water supply, but secondarily a cannabis grower looking for premium cannabis from healthy plants needs to prioritize the soil and the soil properties important to cultivation. A good indicator of soil health is the natural vegetation growing before planting or preparation. Moss, for instance, may indicate a damp, shady condition with poorly draining soil and not desirable for cannabis cultivation. Trees often indicate adequate soil depth for cultivation, while only grasses may indicate just the opposite. Other natural factors to consider in soil management include the wind and erosion and important when selecting a site for cultivation.

SOIL EXAMINATION AND TESTING

When determining the soil properties for your selected grow site, dig to a depth of at least thirty (30) inches. Tap roots from cannabis plants may go deeper, but thirty (30) inches is the range for most root activity and provides a useful sample volume. A post-hole digger is a great tool for soil sampling since it keeps the hole uniform and makes the soil removal convenient. Examine the excavated material for structure and texture, organic material content (humus) and color. As an initial first step if you do not plan to test the sample, take a small amount to your local nursery or garden center for examination. If near your planting location, these resources are often familiar with proximate soil conditions and can offer expertise in achieving desired conditions for horticulture success.

Chemical analysis by a soil laboratory is a valuable method to determine exactly what characteristics are present and what solutions are available to correct any deficiencies for optimal plant growth. Testing is not expensive and an excellent way to know levels of critical nutrients and micronutrient in the soil. Labs will often suggest amendment programs for agricultural use along with a bioassay of your soil sample. It is a good idea to find your soil laboratory service prior to digging for your sample. Retain a portion of your excavated material for laboratory analysis, according to the requirements and suggestions of the soil testing lab, as procedures vary.

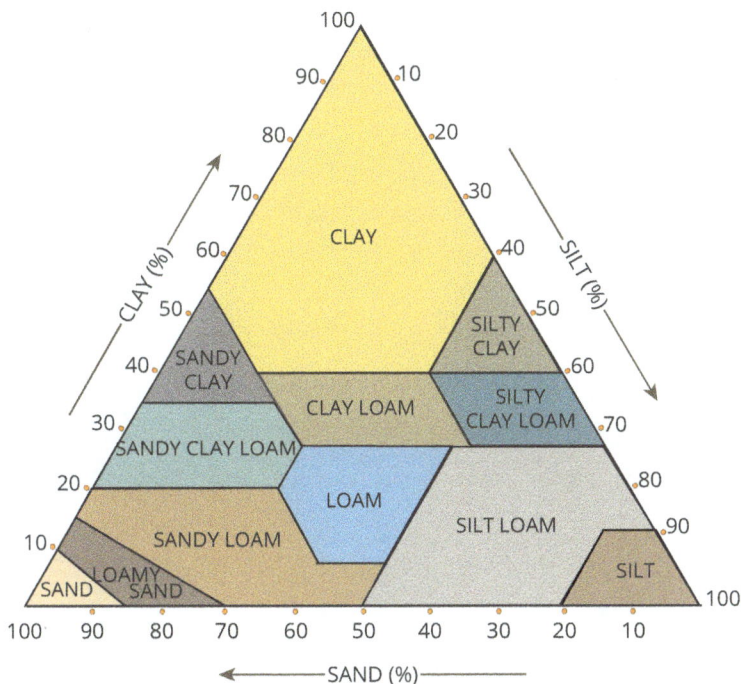

The United States Department of Agriculture (USDA) shows soil types by percentage composition in a graphic widely used by soil scientists and redrawn here to demonstrate the mathematical concept.

A nutritional deficiency is a common problem in cultivating marijuana, whether in field soil, potting soil, or soilless media. When seeking to produce premium grade marijuana, nutrient levels must be ideal, so any deficiency is unacceptable. Laboratory testing is an efficient method of determining the exact levels of both beneficial and harmful contents of any growing material employed by many professional growers. Establishing a working relationship with a soil testing laboratory is an important consideration for both efficiency and cost. Your local agriculture extension office may list testing labs in your area.

While the exact percentages are unknown, most commercially grown cannabis is in containers of some type, including raised beds, instead of planted in the ground. Quite simply, hydroponics and soilless substrates offer better control, cultivation convenience, and higher crop yield than field-grown stock for both outdoor and indoor gardens. That trend is changing as legalized cultivation makes outdoor growing less risky and farmers venture into the field for sun-grown crops where the seasonal conditions allow it and in-ground planting is feasible.

Even so, comprehension about the fundamentals of natural soil and the role it plays in vital plant processes is useful to any premium cannabis grower who can use those principles to succeed in today's market and become inventive for tomorrow's. If you want to be a great marijuana farmer, learn as much as you can about soil; it is your most influential partner in cultivation.

Knowing the vital role soil plays in crop success, you cannot get too much guidance on the subject and a great resource to lead the way is your local Farm Bureau or Agricultural Extension Office. While they may smile at the plant you have in cultivation, most of these organizations will give direction on where to get help locally with soil. That is important because soils vary so much by location, you need expertise on your regional conditions. In addition, Farm Bureaus and Agricultural Extension Offices are a wealth of contact information for vendors, testing labs, and other resources that can help you make your soil ideal for premium cannabis.

The primary goal of a farmer is to give crops the best conditions to grow from start to finish. In cannabis cultivation that begins from the ground up, by keeping the substrate ideally suited for optimal root development. The result is healthy plants and higher yields.

If your growing medium is ideal for cannabis cultivation, it should go beyond just providing support for your plants, it should also give air, water, temperature regulation, and nutrients in levels that are most beneficial to plant health and increased growth for maximum yields. Growing mediums should also protect your plants from harmful toxins that effect growth and product quality, like potency, density, flavor, and aroma. The animal life and natural processes that occur in soils and grow mediums have a significant effect on the quality of your crop, so the first step in becoming an expert in a premium cannabis growing plan is to understand that desirable soils and soilless mixes are living organisms.

PHYSICAL COMPOSITION OF IDEAL SOIL

What you want in any substrate for growing marijuana is a way to provide a healthy environment for roots to absorb water, oxygen, and nutrients. This environment and how you maintain it is important from seed to harvest; neglect at any stage will not produce a premium grade cannabis flower bud, no matter how much fertilizer or "special formulation" you pour on it. Plants grow and develop at continual rates and while may appear forgiving to errors in cultivation, if you want the best crop, monitor the soil regularly. So, one technique in achieving ideal soil is to decide to check it at least once a week and more often if issues appear. This applies to plants in the ground, in a planting bag, in a container, or growing bucket. A calendar or check list insures you stick to the schedule.

Best Practice: *Examine your living plants daily and your living soil weekly.*

Look for any abnormalities or changes in color or appearance of the soil within the drip line of the plant (border of plants canopy). Salts in fertilizers; for example, often exhibit in a white crust on the soil surface. Mold growth may indicate too much water or worse; a root disease. Pests like fungus gnats may populate growing mediums and appear as adults above ground while their larvae are developing at the root zone, feeding on roots, stem tissue and organic matter. Early detection can help minimize severe damage, especially to young plants, and serves as a good example of why regular examination of the soil or growing medium is critical. Any obvious changes in the appearance of your soil should get your attention for examination and identification of any pathological condition for immediate correction.

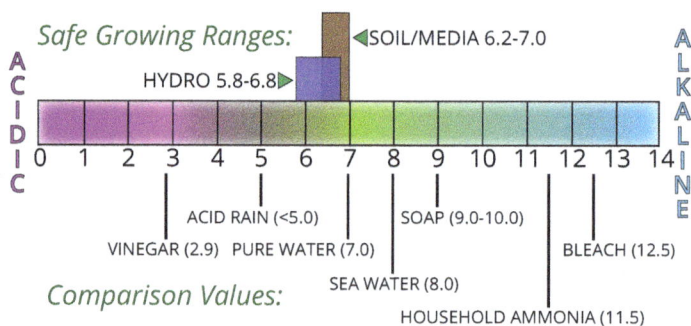

pH scale with common values and safe growing ranges.

The **pH** of soil or growing medium is one of the most important factors in cannabis cultivation, and is highly susceptible to irregularity from the application of water with an incorrect pH, or from nutrient solutions that interact with the growing medium. Meters with soil probes are helpful to measure the pH, but more accurate methods using chemical analysis give better results. Hydroponic stores carry analysis kits useful for this purpose. Some use litmus paper, but digital meters for measuring pH of water are very accurate and a suggested tool for successful cultivation in any method. Using water with the correct pH level is a practical method of maintaining soil health for your marijuana plants, especially when seeking premium grade crops, but regular monitoring is just as important.

Plants need **moisture** for growth, but correct levels are often unknown by new growers. Too little water and your plants will stress from dehydration and too much; they will drown from a lack of air in the soil. So, what is ideal? There are no definitive answers as growing conditions like temperature and humidity in a natural environment, indoors or in a greenhouse varies widely, but a simple way to tell whether there is enough moisture in the soil is to take a small clump in your hand and if it sticks with gentle pressure, it is correct. If it easily falls apart it may be too dry or if it stays clumped like a ball of clay, it may be too wet. Look to your plants as good indicators of water levels in the soil; leaves wilting is the first sign of a moisture issue, either too much or too little and should cause immediate concern. Note that slight drooping of leaves, especially large ones at the end of the day or light cycle are a natural cooling mechanism of the plant and are not an indicator of incorrect moisture content in the growing medium.

Experience pays off when growing cannabis in many ways and one of them is judging when to irrigate your plants. Outside growers must monitor water requirements based on climate conditions and inside growers utilizing hydroponics must monitor reservoir levels; both to prevent dehydration to a dangerous or inadequate level.

Capillary water (also called capillary moisture) is water in the soil held in place between the particles by surface tension. This force moves the water horizontally, while gravity pulls the water downward. Field Capacity is a term in agriculture that measures the water from rainfall or irrigation against these forces.

Field Capacity is a term to describe the moisture held in the spaces between soil particles after excess has drained away in a set time, usually between twenty-four hours and two to three (2-3) days. Also, important is to be aware of the Permanent Wilting Point (PWP). PWP is the minimal amount of moisture in a substrate required for a plant not to wilt. You obviously do not want to come close to PWP as it is very risky to plants in a vegetative stage. Some recovery is possible from slight wilting, but it can be damaging none the less. Adequate moisture must be available to growing roots and at the right ratio; below Field Capacity and above PWP.

Air is a necessary ingredient in soil or growing media for plant development and required for growing marijuana. Farmers till soil to mix in cover crops, amendments, and fertilizers, but also to aerate the soil. This is an important step when planting; both for in-ground and container cultivation.

The aggregate particles in soil arrange in various densities. These arrangements are useful in soil identification and for determining soil use. Dense, or compacted soils lack pore spaces needed to hold water and air; both critical for healthy root and plant development. Tilling, or cultivating, is a method of loosening these soils and thereby adding pore spaces. Subsequent watering and a process of natural settling will reduce these spaces over time, requiring more tilling. It is important; however, to avoid disturbing the soil near growing plants. Cannabis plants grow extensive root systems, either from a seed or clone, and some of these are near the surface.

Turning, tilling, or aerating soil or growing mediums too frequently may cause more harm than good. Beneficial organisms within the growing medium have complex life histories, and disruption of habitat (soil/media) is possible when these organisms are most active; usually during the growing season when soil is warm and moist. Instead of tilling for weed control, for instance, use organic mulches instead.

The fourth component of ideal soil is the **humus** content. Organic material is a part of the topsoil horizon and necessary for soil tilth. It is difficult to define the complexity of humus, but essentially it is the result of decomposing organic matter. Humification can be a natural process or the result of composting, but it is chiefly responsible for the fertility of soil, both in a physical way (texture, structure) but also in a chemical way, like increasing oxygen content among the molecules. The result of the humification process is a mix of chemicals from plants, animals, and microbial activity, and has many functions as well as benefits to healthy soil.

Consider humus as the life force of soil because it not only helps to retain moisture by increasing the micro porosity of the soil parts, but it also helps in the suppression of soil pathogens that lead to plant disease. Humus also increases the cation exchange capacity, making it able to store nutrients in the soil.

Part five of ideal soil for growing cannabis is the **nutritional content** required for healthy plant growth. In native, non-cultivated soil, address this aspect at planting time to correct deficiencies, but in all soils, monitor it closely as plants grow and absorb nutrients. If your goal is to cultivate premium quality cannabis, the application of fertilizer is the primary mechanism to enhance the growth of your plants. In the today's world of hybrid strains where nutritional demands are high, this is not an option.

Classification is by elemental content; the nutrients required for healthy marijuana plants and compounds containing these elements are the basis for all fertilizers, both organic and inorganic. Plants consume macro nutrients in larger amounts than micronutrients and macro nutrients are present in plant tissues in larger parts-per-million than micronutrients. Micronutrients are very important; however, because they are present at locations of enzymes responsible for plant metabolism.

PLANT NUTRIENTS

•Macro nutrients

   Nitrogen (N), Phosphorous (P), Potassium (K)

•Secondary Macro nutrients

   Calcium (Ca), Magnesium (Mg), and Sulfur (S)

•Micronutrients

   Copper (Cu), Iron (Fe), Manganese (Mn), Molybdenum (Mo), Nickel (Ni), Zinc (Zn)

The sixth and maybe most overlooked part of an ideal soil for growing cannabis is the **beneficial living organisms** within it. For if the soil is healthy and perfect for growing cannabis, it must be alive and working as an ecosystem in harmony. Fortunately, there are a wide variety of resources to obtain beneficial soil organisms, widely marketed individually, part of nutritional amendments and even in soil mixes designed for optimal growth.

BENEFICIAL ORGANISM POPULATIONS (DIRECT)
Some soil organisms have a symbiotic relationship with plants, including:

•Fungi

Mycorrhizal fungi interact with other organisms in the soil, in the root, and in the rhizosphere. These interactions can inhibit or stimulate however, mycorrhizae are useful in cannabis agriculture for both control of harmful pests and for growth stimulation. In addition, mycorrhizal fungi help to improve overall plant health and help to reduce stress from the environment.

Some soils and amendments inoculated with mycorrhizal fungi and spores of mycorrhizal fungi are available in different sizes by several manufacturers supplying hydroponic and garden center stores. (Note: Experimental studies on mycorrhizal fungi products and their benefits are currently non-conclusive and thus not suggested for broad use by many researchers. Cannabis farmers, however widely use this beneficial organism.)

•Bacteria

Soil-inhabiting bacteria are present in large numbers in productive soils. Some are beneficial and serve to decompose organic materials, and help to keep nutrients at the root zone. Inoculates are available in liquid form that contain not only mycorrhizal fungi populations but also beneficial bacteria. Easy to dilute for application, these formulations are good for hydroponic media, soilless, and native soil applications.

•Nematodes

Most species of nematodes (small unsegmented worms) are beneficial, and those sold in garden stores will be. Beneficial nematodes feed on bacteria, fungi, protozoa, and other nematodes like root eating nematodes.

SOIL HOSTING

Soil biologists encourage gardeners and crop cultivators to nurture and protect existing communities of beneficial soil organisms instead of introducing new colonies of outside organisms through purchased products. Methods of hosting beneficial organisms include:

•Add organic matter to the soil.

•Irrigate efficiently to achieve uniform dampness.

•Till seasonally or only when necessary.

•Minimize pesticide and herbicide use.

•Avoid plastic sheeting on soil surfaces.

AN IDEAL ENVIRONMENT FOR LIVING SOIL

The relationships between the biological, chemical, and physical components of growing media create an environment of living organisms. So, the ideal soil for growing premium grade marijuana is a mix of the right physical ingredients combined with the correct conditions that work together in creating growing media best suited for optimal plant growth.

As a grower, you can attain maximum yields if you think of the substrate as a significant part of premium cannabis production. For good health and vigor that will yield a desirable crop, the components of your growing media require conditions that will be the most beneficial to the natural processes that benefit the changing needs of a cannabis plant. While it may be challenging to attain perfection when working with living materials (soil/media and plants), four physical factors play a significant role in maintaining good soil health for growing cannabis.

•Soil Type

Ideal soil type, natural/native: Loam

The perfect blend of Clay, Silt, and Sand for cannabis crop production; provides excellent porosity with a high capacity to hold water and available nutrients. Amend native soils to structure resembling loam.

Ideal soilless medium: Coco Coir; used alone or mixed with other substrates.

This component of growing medium is versatile for both indoor and outdoor uses. Variation of the quality of fibers occurs among producers and reflected in price. Look for products that have low salt content (rinsed in rain over several seasons is best), and long fibers. Most manufacturers will label their products to reflect these qualities.

•Soil Temperature

Ideal soil/media temperature (Range): 65° F. to 70° F. (18° C. to 24° C.)

Ideal irrigation water temperature or nutrient solution temperature used on soil/media (Range): 68° F. to 72° F. (20° C. to 22.2° C.)

The temperature of any growing media plays a critical role in the health, or lack thereof of the cannabis root system. Your management goal should maintain an optimal temperature without drastic swings, either warmer or colder, throughout both the vegetative, and bloom cycles. Here are some simple ways to keep your soil temperature correct and static.

•Avoid black containers in hot sun; cover or tarp with reflective or white materials.

•Use water that is close to the ideal soil temperature range for irrigation.

•Elevate reservoirs from floors to keep the air flow even around them. (Concrete serves as a heat-sink and warms slowly).

•Monitor soil or grow media regularly for Moisture, pH, and Temperature; adjust accordingly.

•Keep soil/grow media from wide swings in temperature.

Temperature extremes like near-freezing or excessive heat alter properties of soil and may harm beneficial organisms within it; especially non-native species often found in quality potting soils. Ice-cold irrigation water for container plants in the hot sun is dangerous for cannabis root systems, just as hot water will harm roots in cold soil.

•pH

Ideal pH (Range): SOIL: 6.0 to 7.0, HYDROPONICS: 5.8 to 6.8

For absorption of nutrients at optimal levels, use the pH range according to the growing method. Each planting method uses different water and nutrient delivery systems with different results in chemical processes. Measure, adjust, and monitor all growing mediums because of so many variables and changing conditions. Amended native soil or soil in container gardens requires a slightly more alkaline pH than hydroponic growing media. Using a slightly more acidic range is an example of a method to offset changes in substrates like coco coir used in hydroponic culture.

•Moisture Content

Water is vital for plant survival, but the proper moisture content (water available to roots) in the growing medium is the key to obtaining a premium crop. Beyond the right amount delivered at the right time, water quality is the other aspect of moisture content in the considered variables. The water holding capacity of soil or grow media and the quality of the water used for irrigation is equally important as they effect a plant's ability to absorb not only the water, but nutrients as well. If you irrigate your cannabis plants correctly, think quality as well as quantity.

Finally, irrigation water and soil moisture are so significant in maintaining an ideal growing medium for optimal plant growth, an entire Chapter covers it. See: Chapter 6, "Water Resource Management: Making Every Drop Count."

The first objective in the pursuit of premium marijuana flower buds is to create an ideal environment for the root system. While you may not reach nirvana at every step of the growing process, a strategy that creates a good medium, maintains it at close to desired ranges of critical factors and then monitors it regularly will achieve the best result.

Substrates Best for Plant Age

There are distinct advantages to using different substrates according to the age of a cannabis plant, particularly at the early stages of development. The premise is to provide the optimal conditions based on the changing needs of plants. Transplanting is not without risk, so a methodical and planned approach is a wise policy using this helpful procedure.

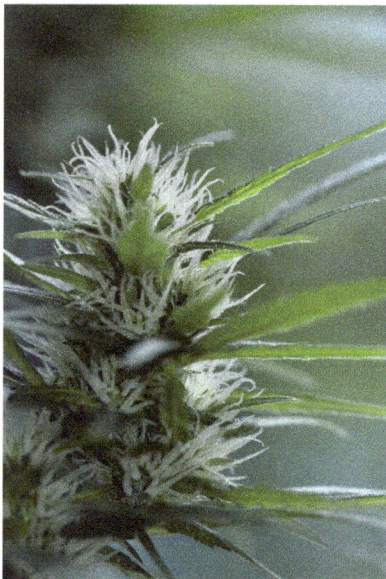

When employing selective substrates, the use of smaller containers in the early stages adds to the benefits of utilizing specific media formulas, affording larger production numbers in smaller spaces. Smaller containers early-on significantly reduce water and nutrient requirements in that there is less substrate to saturate and less root space to satisfy; an added payback for this practice of promoting faster, healthier growth.

An ideal progression starts with seeds inserted into small peat pellets, one (1) to two (2) inch cubes of floral foam designed for propagation or into conditioned rock wool cubes, and covered with one-quarter inch of vermiculite. Clones (cuttings) with a rooting hormone (gel or powder) applied to the cut end do well when inserted into floral foam or rock wool cubes; a common practice in propagation. When the seedlings are a week old or when the clones show roots on the cube walls, it is time to transplant the pellets or cubes into either four (4) inch nursery cups or preferably, one (1) gallon nursery containers. For both the cup and the container, use a specialized substrate containing an ideal mix for young plants, with a light fertilizer component. A four (4) inch rock wool cube is suitable in hydroponic cultivation for this step and stage of control.

There are several soilless mixes on the market ideal for seedlings, young plants, and for complete growing of all ages. Some growers prefer to keep the base substrate the same from the one (1) gallon container to the permanent container and simply amend the mix for young plants with added perlite or vermiculite. This method is effective in reducing shock from a transplant and makes conditions conducive for roots to easily move into new material.

When the plants establish a developed root system and reach about twelve (12) inches high, they are ready for placement in a permanent location and larger container for completion of the crop until harvest. The media for this last step in planting must be applicable for the cultivation method, with improvement as required.

Applying this stepped procedure and using a substrate according to plant age are a useful technique that assists in catering to the changing needs of healthy plants. Although it requires some additional labor, the benefits of reduced water and fertilizer outweigh the cost; more important the plants receive optimum substrates, promoting vigorous growth and robust blooming.

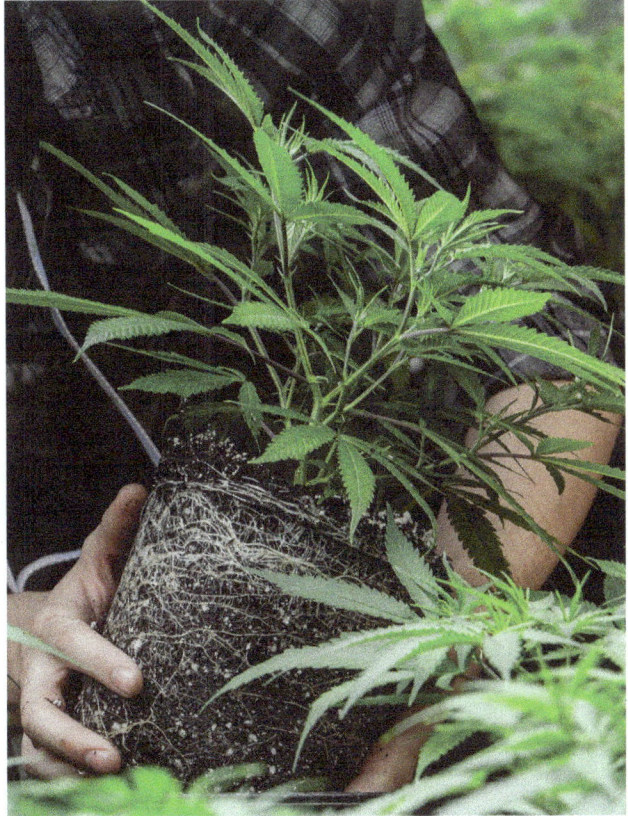

The root systems of cannabis plants, though unseen and often overlooked, play a vital role in the health and vigor of crops in cultivation. Growers seeking top quality yields can use nurturing techniques to ensure what is below the surface is ideal to benefit production results on the top.

Soil scientists and horticulturists estimate that roughly eighty (80) percent of plant disorders are from soil and root issues. Maintaining a healthy and vigorous root system is a primary goal for growing cannabis and for top-shelf results, imperative for success.

The basic function of the root system is to provide stems and leaves with water and dissolved minerals. The process requires that the roots move into new regions in the soil, or growing media, so in simple terms, making it easier for the roots to find oxygen, water, and nutrients provide the conditions for healthy growth and abundant yields.

Note that root growth and metabolism are dependent on photosynthesis in the leaves, so the significance of a what goes on above the ground is just as vital to the roots as what goes on in the soil. Like the seed; however, it all starts underground. The beginning of the vascular process of moving water and nutrients occur in the roots and have a huge effect on the size and vigor of plants. Any approach to growing premium grade cannabis must consider the importance of the root system and how to keep it healthy for optimal plant performance and the resulting quality crops.

The primary functions of the cannabis root system, developed from seed, or a cutting (clone) will anchor and support the plant; absorb and move water and minerals; store carbohydrates, sugars, and proteins (from the photosynthesis process). The roots have no commercial use, and contain no THC.

Cannabis plants grown from a seed have a taproot system with one main root and smaller lateral roots. Taproots are important plant adaptations that assist in search of water. Plants started from cuttings, or clones, have a fibrous root system of similarly sized roots. Each type of root system offers advantages and disadvantages to the commercial grower, influencing the selection of the best suitable propagation or most useful cultivation techniques.

ROOT MORPHOLOGY

•Epidermis: Outer layer of cells

•Root Hairs: Unicellular extensions of epidermal cells. Absorptive and the major location of water and mineral uptake. Root hairs are vulnerable during transplanting; they are extremely delicate and dry out quickly. Clones and young plants generally are susceptible to transplant shock from dehydrated root hairs, so take care to avoid over exposure to drying conditions.

•Endodermis: A single cell wall between the cortex tissues and the pericycle.

•Cortex: Primary root tissue.

•Pericycle: Branch roots begin in this layer of cells.

•Vascular System: The xylem and phloem tissue group.

•Phloem: Tissue that transports products of photosynthesis from the leaves and stems throughout the plant, including the roots.

•Xylem: Tissue that transports water and minerals up from the roots to the plant.

ROOT ZONES

Maturation: Arterial region of the root transport water and nutrients.

Elongation: Region where new cells enlarge.

Meristematic: Area of root tip and root cap.

Root Tip Meristem: Area of cell division and elongation, at root tips behind the cap.

Root Cap: Thick walled cells push through soil like a metal thimble and protect meristem tissues.

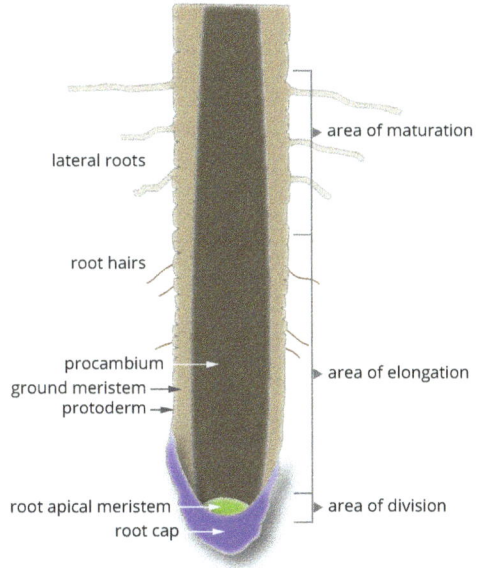

lateral roots

root hairs

procambium
ground meristem
protoderm

root apical meristem
root cap

area of maturation

area of elongation

area of division

ROOT SYSTEM SPREAD AND DEPTH

It is challenging to accurately predict the root spread of any specific plant as cultural and conditional variances play a huge role. Even using various formulas to estimate the root growth under favorable conditions, plant genetics, and soil properties will largely determine how deep a cannabis plant will grow and how wide it will spread.

The larger the planting hole (both depth and width) or the larger the pot, the better. It may or may not lead to a larger plant, but the more space of optimal soil or media you can provide; the more area will be available for extensive root systems to grow and flourish. More roots really do mean more fruits.

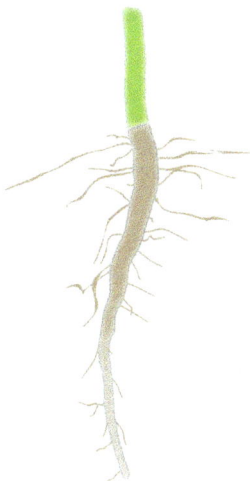

•Tap Roots

When a cannabis plant grows a tap root from a seed, that tap root will go in the direction of least resistance to get the water and minerals for metabolism. Typically, this is downward, but an examination of multiple root systems of the same species grown in the same soil will demonstrate that root systems are unique and variable.

In a container like a planting pot of 10-gallon size, the tap root often heads to the bottom first and then wraps at the base as branch roots fill the balance of the container during the vegetative cycle. Planting depth or width alone does not affect the size of your plant; rather it is a combination of available room to grow above and below ground, the genetics of the plant, the growing conditions, and the ability of roots to absorb water and nutrients. Plants grown hydroponically with a constant supply of water and nutrient solution require less space than that of an outdoor grown plant in the ground or a container; for example.

Clones will produce root masses that typically grow evenly through the substrate. With attentive watering and fertilizing, these fibrous roots may not go deep enough at a fast-enough pace to provide adequate support for rapidly growing plants as they mature. Use staking or netting for clone cultivation, for both containers, and plants growing in the ground.

The growth habit of cannabis hybrids is genetic, such that some strains will grow deep with spreading root systems that may be more tolerant of lower oxygen levels than those strains that grow more shallow root systems; programmed from parent genetics. For instance, indica dominant strains often have more shallow root regions than sativa dominant strains that typically tend to grow deeper with less lateral branching. This is a generalization to illustrate the point as even the same strains from the same parent stock will vary in root development.

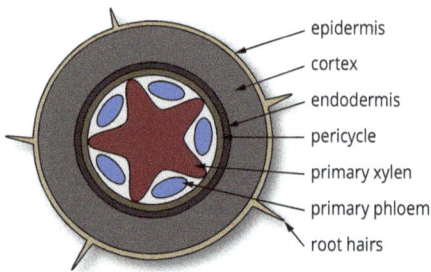

epidermis
cortex
endodermis
pericycle
primary xylem
primary phloem
root hairs

Enlarged cross section of a cannabis root.

Soil types also play a big part in the depth and spread of cannabis root systems. Greatly influenced by water penetration and content, compacted soils, or soils heavily with clay will cause roots to remain near the surface where oxygen is more readily available. In dry or sandy soils, roots will spread and grow deeper in search of water.

Additionally, beneficial fungi in the soil form a symbiotic bond with roots that help in water and nutrient uptake and will influence the depth and spread of cannabis root systems. Beneficial fungi like mycorrhizae enhance the root's contact with soil, although the full range of benefits is not known. Hence, cannabis growers often prefer potting soils and amendments enriched with mycorrhizae, or will add it as a supplement to their planting mix.

POST HARVEST EXAM

Monitoring critical aspects of cultivation during a growing cycle is important for success, but an assessment after the completion of a growing cycle can reveal useful information for future projects. Looking at the root systems of plants after harvest is an effective method of evaluation.

To get a good sample, do not pull the remaining plant trunk from the soil or media. Instead, dig up the root system and remove the soil or growing media from it carefully by shaking. If removing the root mass from a container is difficult, let it dry thoroughly first. As it dries, the root system and substrate will shrink for easier extraction.

Look for growing patterns of the roots, densities, colors and any abnormalities, or evidence of pathogens, or pests. A microscopic examination by yourself or a laboratory is a good practice for identifying disease or root health issues as reference for future cultivation or required sterilization after a given crop. The examination is not expensive and a valuable process for a serious grower.

Catering to the needs of healthy root systems begins with careful preparation of the medium in the planting hole or container and continues with regular maintenance throughout the life of the cultivated cannabis plants, right up until harvest. Considering the roots each time you plant or perform regular maintenance like irrigating or fertilizing your plants is a good practice in premium cannabis cultivation and one that will serve you and your plants, very well.

The Rules of Nature

The most difficult part of describing cannabis and how to grow it, or any plant species for that matter, is explaining all the variables that influence every part of the growing process. Nature offers the infinite uniqueness of all life as the norm, but for a production farmer, that fact is not so great. A cannabis grower needs uniformity and predictability to produce consistent results, especially in a crop like marijuana, where "oddballs" are unwelcome at the party.

A marijuana farmer benefits from accurate information that can help develop procedures for expected results. A reasonable expectation that if you grow your plants a certain way, and follow all the suggested best practices of cultivation, the resulting crop should be outstanding. In theory, that is the goal; however, anything can happen to change the most detailed plans. Planning for what might happen is part of farming cannabis; the best way to do that is to realize that nature rules the day.

Written to better understand the cannabis plant and best practices that can work with natural processes, this book describes how to help direct the desired result of cultivation. It is simply a road map of where others have traveled to farming success; where you go with it the information is up to you. If you use even some of it to help your operation, then you would have discovered that working with the rules of nature is truly the art of growing premium cannabis.

Even seeds from the same parent plant develop natural variations; emerging roots from peat pellets are an indication these seedlings are ready for planting.

Sun Grown Goodness

Some cannabis strains perform exceptionally well in the sun, and some are best suited for indoor growing. If your climate zone makes outdoor growing possible and you have the land, there are distinct advantages to sun-grown cannabis, not the least of which is a higher price for the finished product. If you plan from that standpoint, choosing an outdoor, sun loving variety may be a good decision.

From an economic angle, growing in the sun eliminates the need for lighting and cooling of those lights, with no utility bills. It makes sense provided your plants will receive at least six (6) hours of direct sunlight per day. Any less than that and you will run the risk of low weight crops and of poorer quality than those grown in the full sun. In Northern California, clandestine gardens planted in the deep shade of forests might appear in the news, but these plants are pathetic producers compared with nurtured plants that bask in the sun and are rarely the projects of savvy growers.

Sunlight is energy to a plant that uses it in a photosynthetic process to facilitate growth. For a cannabis grower, that light energy influences flower production in quality and weight, so it is a determining factor in choosing where to grow and what method of cultivation is ideal for that location. Since outside growing is dependent on the growing seasons and windows of opportunity for starting and completing a crop, farmers may only have one production cycle per year; another thing to think about in planning for profit and best yields.

If you do grow in the sun, make certain to use that as a selling point to your customers in the distribution chain. Consumers will often look for sun-grown varieties and correct labeling is very beneficial.

Chapter 3
Cultivation Methods: How You Grow is Everything

As we learn more about cannabis cultivation with new ideas and improved technology, the marijuana farmer has several ways to grow their plants. Each method has both advantages, and disadvantage when compared on chief criteria and so no single way produces the best crop result for every situation. The goal in each method is to provide plants with the water, air, and nutrition required for growth, based largely on where cultivation occurs (indoor, greenhouse, outdoor) and how; by aeroponics, hydroponics, deep water culture, growing in containers, fabric pots, or in the ground.

If you are planning a new cultivation setup, or are thinking about a change from your present method, consider your entire project, a growing *system*. No single part of the farming process, despite project size, is independent of another part. A plant requires a support structure to grow and supply it with life essentials of water, nutrition, and air. The growing media, irrigation, fertilization, and environment all work together in a system, and refining any part of it makes it better for the whole process and ultimately the plants you are growing.

When selecting a growing method, it is also the time to choose whether to grow your marijuana plants organically, or use synthetic compounds. Both have merits to consider according to your chosen cultivation method.

3 | a. Define a Growing System
The Customized Marijuana Farm

Scrutinizing your cultivation method is something you will do always as a grower of premium products. When starting, your way of growing is a very important consideration effecting your wallet as well as your crop since it determines how you will create your ideal garden. If your goals have changed or you are thinking about an upgrade, there are many choices to consider for an awesome farm that fits your needs and aspirations.

F rom an agricultural perspective, cannabis is a short cycle crop. As a fast-growing plant, the most suitable substrate for cultivation depends on the method of nutrient delivery and the route or destination of any solutions used in that delivery. Together, the substrate, irrigation and nutrient delivery form a cultivation system.

SOIL BASED SYSTEMS

-Direct in-ground planting

-Potting soils (Pre-fertilized soils or amended soils)

-Soilless mediums

Inorganic

Gravel, sand, clay pebbles, perlite, vermiculite, rock wool

Organic

Peat, saw dust, bark, rice hulls, coconut coir, forest material

•Nutrient Delivery for Soil Based Systems

Fertilizers and supplements in dry form, mixed with the medium, or in liquid solution applied to medium surface for penetration to root zone. Soils and soilless media require the regular addition of nutrients for best crop yields. Container cultivation is typically a drain to waste process as described for Hydroponic Systems.

HYDROPONIC SYSTEMS

-Closed (Re-circulating)

Plants are in direct contact with a nutrient solution that recirculates. (Periodic change outs required.)

-Open (Drain to waste, Run to Waste)

The nutrient solution not recovered for immediate use, sometimes collected for processing including filtration and then reused. Usually waste; may drain to hard surfaces or surrounding soil.

•Nutrient Delivery for Hydroponic Systems

Cannabis strains grow successfully indoors or in a greenhouse using different hydroponic methods, with or without growing media depending on the technique used.

-Aeroponic: Nutrient solution provided directly to plant roots in a mist or aerosol

-Ebb & Flow (Flood & Drain): Plants in growing media regularly flooded with solution that drains to a reservoir below

-Semi-hydroponics or Passive Sub-Irrigation: Roots grow in a shallow solution of nutrients, usually supported in media like expanded clay, coconut coir, or both

-Deep Water Culture: Plant roots suspended in an oxygenated nutrient solution

Optimal Cultivation Conditions
Targets for Maximum Growth and Vigor

DURATIONS:	Weeks 1-4	Weeks 1-3	Weeks 4-6	Weeks 7-9	Weeks 10-12	Last 3-7 days
FACTOR	**Vegetative Phase**	**Flower Set**	**Flower Stacking**	**Flower Maturing**	**Ripening**	**Flushing**
Day Temp. (°F)	80° - 82°	76° - 80°	72° - 76°	72°	68° - 70°	62° - 64°
Night Temp. (°F)	70°	64° - 68°	60° - 64°	60°	58° - 60°	58° - 60°
Day Humidity (rH %)	60 - 75	55 - 60	50 - 55	40 - 45	30 - 35	25 - 30
Night Humidity (rH %)	70	40 - 45	35 - 40	30 - 35	10 -20	10
CO_2 (ppm)	1000-1500	1200 - 1400	900 - 1100	600 - 800	300 - 350	300
Water/ Solution Temp. (°F)	68°	68°	68°	65° - 68°	60° - 62°	60°
pH: Water/ Solution/ Media	6.1 - 6.3	5.9 - 6.0	5.8 - 5.9	5.8	5.6 - 5.7	5.7
EC: Water/ Solution/ Media	0.2-0.6	0.8 - 1.0	1.0 - 1.3	.9 - 1.3	0.5 - 0.7	0

Durations are range estimates; variations are likely according to strain and cultivation method.

Environmental conditions that are optimal can be a useful guide when choosing a cultivation method since they show what ranges are ideal and help determine the practicality of how and where an operation can function properly; essential for budgeting costs to setup and run. This chart helps to plan and provides known targets for each category based on the changing needs of a plant by stage of development for a typical hybrid strain that requires ten to twelve weeks in the bloom phase of development.

Note: Values are from multiple sources, widely known and in common use by growers in various applications; however, each plant strain is different and may prefer unique conditions not indicated by this chart. Also, indoor growing affords the greatest opportunity to control the environment, greenhouse cultivation some control, while outdoor cultivation is the subject of natural weather conditions with little control.

SYSTEM CHOICE

Choosing the best system is an individual assessment for each grower based on several variables. The system selection should consider a grower's experience and knowledge, balanced with cost, facilities, and productivity goals. Expansion with some modification of a system is practical as production increases, however changing cultivation systems altogether can be expensive and time-consuming. Careful selection of a system at the start of a farming operation is the most prudent step in growing good quality cannabis.

COMPARING SYSTEMS

•Start with the Cost

The economics of cannabis cultivation will likely play a big part in the selection of your growing system. The hard reality is that you cannot easily change from one system to another without incurring waste and unnecessary expense, so it is very helpful to think about your initial setup costs ahead of time. Developing a simple computation that includes everything you will need to grow your marijuana plants is a smart step on the road to optimal production. During a growing season or cycle, your focus should be on your plants; not surprise expenses that can be distracting, or worse, hurt your plants and ultimately your growing plans.

•Price-Per-Plant Planning

Use a method of cost budgeting called P-P-P; price-per-plant for estimating expenses. Unlike other crops that may grow with many plants per row, most cannabis farming of hybrid strains is akin to producing an elite commodity in limited numbers. So, the best way to plan your farming budget is to figure what each plant will cost to grow to maturity. This formula is applicable to all sizes of cultivation gardens, unless you are using a row system in the ground.

•Sample Formulas

10 Plants, Outdoor, Growing in 10 gallon containers

(Seed/Clone Cost) + (Container Cost) + (Soil or Soilless Medium for that container) + (Nutrient supplements for vegetative and flowering phases) + (1/10 X Total of all other cultivation expenses for the season) = Price Per Plant

10 Plants, Indoor, Growing Hydroponically in 10 gallon containers

(Seed/Clone Cost) + (Container Cost) + (Soilless Medium for that container) + (Nutrient supplements for vegetative and flowering phases) + (1/10 X Total of other cultivation expenses for the season) + (1/10 X Utility cost to run lighting, pumps, heating, cooling, etc.) = Price Per Plant

Cultivation expenses as used in these formulas might include facilities expense, lab fees, delivery charges, etc.; include anything that costs you money to grow your plants. Also, the formulas do not consider startup expenses. For outdoor growing, these costs may be minimal while indoor growing requires lighting and climate control equipment that may not be inexpensive. In a farming business sense, factor these expenses into what it costs to grow your plants over the life of their use.

CONSIDER HIGH PRODUCTION METHODS

Just as there are many desirable strains to cultivate, there are many physical ways to grow cannabis plants for optimal crop production. Each requires different amounts of physical space depending on location, inside or outside.

Using a system called Sea of Green (SOG), tightly spaced single stemmed plants produce quickly because of the intense technique. In SCROG (Screen of Green) systems, plants grow through netting placed horizontally to support short, heavy stems. In both commercially popular methods of intense cultivation, farmers utilize various beds or containers, substrates, and irrigation techniques for mass production of rapidly maturing plants.

Alternatives that produce well includes large fabric pots that farmers use both outdoors and in greenhouses. The larger the pot of soilless medium, the larger the total weight of finished bud is usually the result. Accurate formulas for size of pot to resulting yield are not reliable because of the many factors that influence plant growth. A sixty-five (65) gallon container has the potential for two (2) to three (3) pounds of finished flowers in ideal conditions; for example.

Black nursery containers, made from recycled materials, are a popular choice of growers for low cost and ease of use. Ten (10) gallon size and larger are the best for individual plants unless using high-production growing methods as described. Drain holes must be located on the sides of the container bottom and not under it, or the drainage will not be adequate for cannabis plants.

It is helpful to cover the sides of black pots exposed to the sun; roots will suffer from high heat on the container wall and the substrate heats to dangerous temperatures if left unprotected.

After removal of the growing medium after each harvest, sterilize the plastic pots with a solution of one (1) part bleach to ten (10) parts of water before installing fresh substrate for a new crop.

MAKE CONNECTIONS FOR ADVICE

If you are starting a new cultivation project or thinking about changing your present growing system, before you make the final decision, connect with a few experts. Some of the best recommendations come from salespeople at the local hydroponic stores. They know what is selling and why, along with the experience to suggest what may be applicable for your specific situation, but the wisdom exchange does not stop there.

If you plan to cultivate marijuana at an advanced level, you should be developing relationships with people who can help you achieve your goals; this includes retailers of products you need or will be using, dispensaries where you might buy seeds or clones and sell finished buds, and most importantly, associations and organizations that support cannabis growers. If you are a successful and experienced grower or as you become proficient, sharing your cultivation knowledge with others is good for the industry, the science, and most of all, for you.

So, when selecting a growing system, think about where you will be growing, how much space you have, what method will meet your production goals, how much you want to spend, and lastly, what do others in your area use or suggest. After considering all that, your next step is to choose the chemical aspect of your method; to grow organically or with synthetic compounds. The following section guides you through that decision; an important one for the composition of your finished flower buds and the market value of your crop.

Substrate Volumes

- 4" planting cup (10 cm)
    - 1 pint (.5l)
    - 1 cu. ft. fills about 52 cups

- 10 gallon hard-wall container; 16"d., 12"h.
    - 1.45 cubic feet (cu. ft.)
    - .05 cubic yard (cu. yd.)
    - 1 cu. yd. fills about 18+ containers

- 100 gallon fabric container; 38"d., 20"h.
    - 13.12 cu. ft.
    - .49 cu. yd.
    - 1 yard fills about 2+ containers

- Raised Bed; 8 ft. by 4 ft., 12" deep
    - 1.2 cu. yd.

WARNING! KEEP PLANTS AWAY FROM TREATED WOOD!
Do not use any wood like railroad ties, telephone poles or pressure-treated lumber in the construction of growing beds or nursery tables and shelving. Systemic action will pull harmful chemicals into your plants from soil or media contacting treated building materials, harming your plants and contaminating the crop.

3 | b. Organic Cultivation
More Than a Labeling Term

The benefits of growing cannabis organically are numerous to the environment, the consumer, and the marketability of the finished product. While organically grown crops typically command a higher price than synthetically grown, the endeavor is not without challenges. A crop labeled organically grown must comply with regulations and meet protocol that excludes certain nutrients and cultivation practices. Thoroughly examine the decision to be an organic grower or not and the proper supplies required before starting a crop or setting up a garden; it will help prevent costly mistakes and waste.

Growing marijuana organically, in the truest sense of the definition, will have benefits and disadvantages you should consider. If you decide to grow purely "organic," it is an all, or nothing proposition in that everything you use in the cultivation process, from seed to sale, must be organic, all natural, and without synthetic components. This includes your growing medium, so choose your method before you select the soil or planting location.

If you have ever attended an organized cannabis event, you might have noticed companies who support cannabis growers with the latest technology, including soils, amendments, fertilizers, and yes, magic solutions. Some are awesome; some are not. In the large assortment of products in the marketplace, it is easy to become confused to the point of either buying more stuff than you will ever use, too many incompatible nutrients, or even those that will cause harm to your plants if mixed with incompatible products. Worse are those that cause harm to the environment.

In the mix of new merchandise displayed alongside the reliable, tried, and true, is an emerging trend to the marketing of "organic" products, including soils, and substrates. Why the effort? In areas where marijuana cultivation is legal, there is an increasing concern over unregulated pesticide use among marijuana farmers, so there is a push by regulators for limits on what chemicals are allowable for cannabis plants. Many believe an "all-organic" approach to cultivation will solve all their problems, but this is not so.

Growing marijuana can be organically, or inorganically, partially, or completely, with varying results, anyway you want to measure them; that is the nature of cultivating plants. What farmers must not do is confuse the methodology with the real issue, and that is contaminants. Since there are different contaminants and various ways your crop can become contaminated, understanding how to avoid stuff you do not want in your plants helps protect your investment of time, labor, and money. More important, it results in a better crop, keeps you safe, protects the environment and helps comply with regulations.

One of many contaminants that are an enemy to your crop, both during growth, and after harvest is from biological agents like mold. There are various times at which microbiological contamination from different sources can occur. During the cultivation, harvest, storage and final distribution of marijuana flower buds and derivatives, harmful microbiological agents like mold, mites or insects can infect your product. While any contaminant is bad, contaminants include unacceptable levels of chemicals used in the growing process; chiefly pesticides and supplements containing heavy metals.

As both medical marijuana patients and recreational users are seeking more tested product, even unregulated dispensaries or retailers are requiring some sort of lab analysis, pound by pound, before purchasing from growers. Organically grown product must also pass scrutiny of what is safe for human consumption. Distributors and consumers want to know the THC content obviously, but they also do not want mold, mildew, germs, herbicides, pesticides, and harmful chemical content in the weed they buy and ingest. How you grow your marijuana, the soil you use, the sprays you employ, and the fertilizers you utilize all have a direct effect on contaminants.

Let us say that you grow responsibly, do not use harmful chemicals, and do not plan to test or have your product tested. You plan, like others, to identify your crop as "organically grown." Just as consumers have increased their spending on organic foods, you know that cannabis buyers are also looking for less harmful substances in their smoke, tinctures, and edibles. From an ecological point of view, growing cannabis organically sounds like the wise thing to do, and from a marketing perspective, it sounds great. Unfortunately, organic in cannabis cultivation does not always produce the safest crop and may have different meanings to different people.

A local grower that I met once at a farmer's market for cannabis, touted his product as "organically grown outdoors." He had signs, labels, and brochures. Although untested, the buyers of his buds likely assumed his product was somehow safer than the man next to him who grew indoors with manufactured nutrients. The organic producer said he used cow manure as fertilizer. No harm there except the manure he used was from an unknown source. It might have come from some polluted barnyard; who knows? Some of the "organic" buds also showed signs of mold and pest damage. The man selling next to him had pleasant looking, potent-smelling buds along with a certification that his product was free from contaminants. Which do you think has more value? Which one would you smoke? Moral of the story: be very careful with labeling. Often regulated, your product label should be as forthright as possible if you expect credibility in the marketplace and long-term success.

When a chemical compound is highly soluble in water, plant tissues absorb it through the vascular system (phloem and xylem). This systemic action will happen with these soluble substances from both pesticides and fertilizers. It is important to begin an organic farm, even in a greenhouse with raised beds, before these compounds are in the earth. Subsequent planting will reflect what was in the soil when roots reach depths beyond the beds or amendments. Since cannabis roots grow deep, this is critical to avoid contaminant issues that will likely make your finished cannabis crop unsaleable.

DEFINITION OF ORGANIC FOR THE CANNABIS GROWER

It is important to recognize that the term "organic" is both an accurate description of products used in growing or production and a misunderstood marketing term. Consumers accept that Organic on a food or agricultural product means that production meets approved methods, without the use of synthetic materials. Actual definitions are far more precise and are especially pertinent for the cannabis grower.

There is a distinction between USDA (United States Department of Agriculture) certified products (Cannabis is not on the list) and OMRI (Organic Materials Review Institute) certified products used for organic production, like fertilizers and amendments.

About the law and labeling food crops, the USDA is very specific in their regulations; shown in this outline cited from public record. The regulations serve as criteria for how an organic cannabis farmer might approach the issue based on law for food crops. An in-depth look at the entire section, available on the Internet, is a useful reference.

Title 7: Agriculture

PART 205—NATIONAL ORGANIC PROGRAM

Subpart B—Applicability

§205.102 Use of the term, "organic."

Any agricultural product sold, labeled, or represented as "100 percent organic," "organic," or "made with organic (specified ingredients or food groups)" must be:

(a) Produced in accordance with the requirements specified in §205.101 or §§205.202 through 205.207 or §§205.236 through 205.240 and all other applicable requirements of part 205; and

(b) Handled in accordance with the requirements specified in §205.101 or §§205.270 through 205.272 and all applicable requirements of this part 205.

Labeling your cannabis product "Organic" may best describe only your method of growing and may be misleading. "Organically Grown" may be a better description, providing you use only fertilizers, soil amendments, and other products that meet the National Organic Standards.

The OMRI is a nonprofit organization that independently reviews products for use in organic horticulture using standards of the National Organic Program (NOP). Their seal on a product label may also include phrases such as:

"This product is listed by the Organic Materials Review Institute"

"Meets requirements of the National Organic Program for use in Organic Production"

"Acceptable for use in organic production"

"Certified for use in organic farming and production"

"Allowed for organic production"

LONG TERM GROWING STRATEGY

Before purchasing any product moving forward, decide to develop nutrient and pest control plans. Planning should include a list of the products you will use throughout the grow cycle; from seedling or clone to day of harvest. Begin your plan by deciding if your plants will grow organically or not. If you opt to grow your cannabis crop organically, here are some practical ways to ensure your supplies and procedures meet criteria for "organic"; those used by organic food crop growers and producers:

1. Use methods that integrate cultural, biological, and mechanical practices;

• Foster recycling of resources

• Promote ecological balance

• Conserve biodiversity

Not allowed: Use of synthetic fertilizers, sewage sludge (many commercially available fertilizers contain it) or irradiation.

2. Find a current list of substances allowed and prohibited for food crops on the USDA National Organic Program website.

If you plan to grow organically, be careful when developing your nutrient and pest control objectives since many products have the label "Organic," but they not approved for organic farming. Examples include Urea and bio solids, and sewage sludge. Manures are allowable, but composted according to the NOP standards.

Each State has different laws regarding fertilizer, pesticide, and herbicide labeling, so choosing the suitable products and methods for your situation should use the same national standards used for other agricultural products. Although not a food crop, cannabis is a consumable product and you should consider that its composition is a subject of regulation and monitoring; eventually, if not already, despite your location, and authority. Laws pertaining to cannabis cultivation also vary widely; however, as we get closer to legality in more locations, regulations will likely rely on national benchmarks.

Despite your cultivation method, begin with a carefully chosen soil or growing medium, and a well-planned nutrient and pest control directive that meets your goals and complies with rules likely to come either at your State, Province, or National level. It will put you in a better position when the time comes for any type of compliance; licensing, permitting and even approval for your operation. Most of all, it will give confidence to your customer coming from the integrity of your properly labeled, quality product.

The term synthetics often brings frowns from diehard organic growers who have been cultivating weed for decades. Some very good quality nutrient systems are non-organic and rely on formulas that are clean with no impurities or heavy metals. Most are pH stable, highly soluble, and produce a very fine crop of top quality and weight. Before dismissing these chemicals entirely, investigate the healthful result from crops grown with non-organic nutrients.

Growing cannabis in artificial environments with synthetic nutrient systems is becoming a practical alternative as choices of equipment expand, just as high-end growers are looking for total control in the pursuit of premium weed. Gone are the days of completely dismissing synthetic formulas as "non-green" since manufacturing has improved by consumer mandate and law to minimize environmental effect. Notwithstanding, consider the advantages and disadvantages to using either organic or non-organic compounds in cannabis cultivation before selecting.

Many synthetic fertilizers and supplements come from the by-products of the petroleum industry. Some common examples are Potassium Sulfate, Ammonium Nitrate, Ammonium Phosphate, and Super phosphate. The Analysis appears on the product label, with an example illustrated in Chapter 4, page 83.

In specialty stores that sell products beneficial to cannabis growers, there is an amazing selection of synthetic compounds. Most work well when used with the other nutrients and supplements within the manufacturer's line of products. Going outside that parameter is risky because of a danger of malnutrition from nutrient deficiency with the wrong combination and over fertilization from too much of a nutrient. Worse, chemical reactions can occur from custom mixing of fertilizers or application schedules, so avoid it. The sales department of any good vendor can provide specific advice for product compatibility, efficacy, and safety.

pH Adjustment Advice: Allow synthetic nutrient solutions in a safe range of 5.6-6.4 without correction.

If growers select products that are food-safe, apply correctly and are responsible about disposal (waste), there are benefits to using synthetic fertilizers and supplements in cannabis production, particularly in hydroponic culture. Since maintaining the cleanest growing system is possible, or a near-sterile one free from harmful pathogens that often thrive in organic systems, the use of synthetic fertilizers is a choice for specialized cultivation that occurs indoors.

Advantages of Synthetics (Inorganic compounds)

- Generic all-purpose types are less expensive than organics

- Predictable

- Reliable

- Concentrated, very soluble

- Precise Composition (Analysis)

- Odorless

- Fast acting; useful in correcting deficiencies

- pH balanced

- Free of pathogens

- Longer shelf life than organics

Disadvantages of Synthetics

- Specialized formulas for cannabis are more expensive than organics

- Usually contain only a few nutrients

- Diluted solutions leach easily

- May leave residuals not approved for cannabis consumption

- May cause build-up in soil or growing medium

- May cause pH imbalance, soil deficiencies

- Long-term use may disturb microbial activity

- Made from non-renewable resources

- Danger of over-fertilization

There are ways to help mitigate the negative effects of synthetic fertilizer use. Mixing the synthetic compound well into the soil around the intended plants or completely diluting a soluble form before use is two ways to prevent runoff that could be harmful to the environment. Store synthetics safely away from children and pets and always follow label directions exactly.

Many breeders and propagators utilize approved synthetic compounds in their work. Using scientific procedures requires both accuracy and a clean working environment, so this segment of the industry commonly uses synthetics for precise results free from pathogens. Added support for synthetic compound use is the accelerated growth obtained; especially useful for "Mother" plants in production during their vegetative phases of development. Synthetic, or manufactured nutrient products specifically designed for plants like cannabis are available from most hydroponic and specialty nurseries; the hard science is behind their efficacy.

3 | d. How to Begin
Starting from Seeds, Clones, or Plants

There are three chief ways to begin a cultivation project; by seed, by cutting (clone) or by small container stock. There are reasons why each may be preferable by successful growers who choose a method primarily because of tradition, strain selection, production goals, cultivation method, economic factors or because they want to control the timing of cultivation phases and harvest.

The start of any growing project in cannabis offers a list of choices that give premium marijuana farmers several advantages. By determining how you will begin cultivation of your crop, you can narrow in on the strain and desired result. Besides starting your own crop, you can also propagate and breed future generations of selective traits. In cannabis cultivation, there is no reason for a single start method and so many farmers use all of them.

FROM SEED

Starting marijuana plants from a seed affords the biggest selection of strains. The downside is that unless you use feminized seeds, a process performed by the seed producer, new plants require sex identification to determine males to remove from cultivation spaces. Male plants have stamens that produce pollen from a part called an anther and will fertilize female plants in the area. The fertilized female plants develop seeds instead of fat, "fruitless" buds. Unless you remove males away from female plants, your crop will become seed makers instead of bud makers with a reduced commercial value.

Seeds are available at trade shows, from dispensaries and online from seed banks. Most of the online seed vendors are not in the United States, so currency exchange may occur in purchasing. Additionally, there are ways to ensure or guarantee delivery, but check out the reputation of any seed company selling exclusively online before completing a transaction.

You will read about all kinds of ways to start your cannabis seeds. The safest, most reliable way is to place an individual seed in a prepared hole about 1/4" - 3/8" deep the size of a pencil within a horticultural cube of floral foam or conditioned rock wool. These substrates are available in various sizes and in sheets that fit in trays. Firmly cover the hole with vermiculite and place the cube under a protective dome, under fluorescent light, and keep it moist, but not saturated until sprouting. Heating mats designed for seed starting are very effective. It is a clean, precise way of germination that will begin in a few days or up to a week depending on temperature and humidity. The cotyledon of the sprout provides nutrient to the seedling; do not fertilize the first five days.

Old timers swear by moistening seeds in wet paper towel placed in a dark place, then when sprouting occurs, placing the seedling in a growing medium with forceps. This works, but it is over handling the fragile sprouts that may lead to plant failure; it also requires the constant checking of the germination of seeds for satisfactory results.

FROM CLONES
(Rooted cuttings; stems are usually six (6) inches high or shorter.)

Clones for flower production are plants established from cuttings taken from female, mother plants. When root systems develop and are adequate to plant in a growing container or outside location, growers remove them from propagation hoods for starting the vegetative stages of development. The cultivation of marijuana from clones is one of the most popular methods for farmers who want reliability and consistency in specific cultivars. The rub is that you must make certain that labeling is accurate and plants are free from disease or pests, or you will have major problems down the road, not the least of which is the contamination of your other plants or facilities.

For farmers in large production efforts, cuttings from female "mother" plants are the most common method of starting their crop. Clones are mass produced by aeroponics in "clone machines" like in the photograph, or in prepared rock wool cubes of various sizes. Horticultural foam cubes that are pH neutral and require no conditioning are also popular for cannabis cloning. Propagators will sometimes use rooting hormones in a powder or a gel to encourage root development on the cut ends before insertion into the media for starting.

Typically, clones begin development with the placement under fluorescent lighting (T-5 tubes are best). Once roots appear on the outside of the starter cubes or foam inserts, clones are ready for planting into soil or soilless media.

Starting from clones results in consistent, uniform, and predictable growth as descendants of desirable parent stock. Gaining familiarity of the cloning process is beneficial to premium growers who choose to propagate their own plants for cultivation.

This is an instance where you must do some homework and a little leg work. Find in your local area, a reliable source that stocks clones from a reputable clone nursery. Search and talk and ask until you are sure that you can trust the source for your future investment of work and expense. Be smart and pay the going rate for quality stock. Saving a few bucks now in your venture is foolhardy, and shopping price is a waste of time. Spend the price for the best clones you can find and reap the bounty later is really the essential point.

When selecting your clones for purchase, you may find them in various growing media like floral foam, rock wool, or a soilless mix containing coco coir; they all provide excellent hosting for rapid root development. If the clones are in rock wool or foam cubes, look for root systems that are emerging from the medium, as well as the health and vigor of the plant. Larger "trunks" (the stems forming the root system) will typically grow better than thin, weak ones but this is often a strain-specific characteristic, so look for overall health as a good determining factor.

FROM PLANTS

(Plants growing in media; from three-inch cups to one (1) gallon. Avoid larger pots for starters when beginning a project for commercial production.)

If you prefer to start with larger plants or it is late in the season, then container stock is a good option. Most sell in nursery stock cups in three or four-inch sizes up to one-gallon size, but up to five-gallon plants are also available. Larger plants get damaged in transit and are slow to acclimate to new environments; most commercial growers prefer to start with small plants. When purchasing any plants, make certain of labeling and that you know the source or at least the reputation of the grower.

Seedlings emerge from rock wool cubes usually within a week of planting.

QUARANTINE PROTOCOL

(Isolate all new stock before planting.)

Clones or small cannabis plants, like any plant material, can harbor pests and diseases. Even stock from reputable nurseries is not safe because of the exposure to other clones and starter plants in the selling environment. To prevent damage to your crop, other plants you already have in cultivation and contamination of your facilities, you must keep your new clones or plants isolated for examination for a period of about ten (10) days. Slightly longer may be better, but 10 days allows observation and treatment for most pests and diseases that usually appear in some stage of development in a life cycle.

Cautionary or preventative foliage sprays containing Neem Oil may be effective if applied at recommended rates for tender seedlings and used as a soil drenching for prescribed pests. Follow label instructions carefully so you do not burn your new clones that remain vulnerable until hardened off to their new growing location.

The method of starting your crop comes down to a matter of convenience, coordination for your operation, and a personal preference. Most farmers are adamant their way is best; however, how well your plants get care during growth has a far larger effect on crop yield.

Care of Seedlings and Clones
the first week

- Keep daytime temperature at 80-82 °F; nighttime temperature at 68-70 °F

- Keep relative humidity at 60-78 %, use vented domes if needed.

- Provide more than 12 hours of light per 24-hour period in regular cycles.

- Avoid exposure to direct sunlight.

- Keep root zone moist, but not saturated.

- Keep indoors for protection, harden off before placing outside.

- Use irrigation water with pH in 6.2-6.4 range.

- Use water with a temperature of 68 °F for irrigation and sprays.

- Keep nutrient solutions weak to mild with an EC of 0.2

- Offer 1000 ppm of CO_2

- Spray with solution dosage for seedlings of preventive sprays.

- Provide gentle air movement to strengthen trunks and reduce fungal pathogens.

Growing premium quality marijuana requires a keen understanding of what the plants require for health and vigor, but that is on top of good, basic farming practices. Without foundational techniques used in cannabis farming, all the fancy equipment and exotic nutrient recipes are a waste. Use sound horticultural methods, tweak them to fit the requirements of the illustrious cannabis plant and you will have a solid path to crop success.

People call cannabis weed in both slang and common usage, but today's hybrids are nothing like a weed. They require great care and nurturing unlike any plant in your garden or farm. Even with the cultural requirements needed to grow a top-quality crop, there are basic concepts and techniques that apply despite the strain or cultivation method. Here, the focus is on the basic care tactics; required labor for best results.

Each day, examine your entire garden, or growing area. Look for any kind of weirdness or abnormality and address it immediately. Whether an irrigation issue, a climate challenge, or visible identification of pests or disease, a daily check will prevent a world of hurt for most problems easily minimized with quick action. This is likely the single most important thing that a marijuana farmer can do in crop care.

VEGETATIVE STAGE

During the first four (4) to eight (8) weeks of plant growth, cannabis cultivars benefit from structural support as they grow and develop weight from stems, foliage, and subsequent flowers. Install early in the vegetative phase since roots make it more difficult to insert staking as the plants mature. Select a method that will be secure, with a way that will not require constant attention or changing. Stakes, netting, welded wire fencing, pyramid frames, etc., alone or in combination provides the strongest mechanisms possible to avoid branch breaking when the flowers mature.

Aside from the mechanical systems needed to help hold them up, your cannabis plants must have protection from harsh climate conditions and pests of many kinds. How you achieve that depends on your specific location and requirements, in consideration of your individual cultivation methods. (Noted because safeguarding from natural factors is a physical labor and material consideration.) This book covers the various levels of climate control and pest management in subsequent sections.

You will read and hear various opinions about pruning marijuana plants. Rule of thumb, do as little as possible. Unlike other type of plants, pruning does not always promote growth and may make your plants more susceptible to damage from irregular growth or disease entry into the cut wounds. Read the pruning guide for more details.

BLOOM PHASE

The principle goal during the bloom phase of development is to provide ideal conditions for the setting of buds and the ripening of flowers laden with resin glands. During this part of the crop production, very little handling of plant material should occur. Stems are brittle and the trichomes that will appear on branches, stems and leaves are vulnerable to damage from contact or touching. Once compromised, these resin glands will not mature and may provide an ideal habitat for fungal infection.

Farmers will often remove the large leaves from ripening bud structures to allow more plant energy for the developing flowers, allow more light penetration into the plant and diminish the labor needed for grooming the harvested crop. When practicing this technique, make certain to remove as much of the petiole as possible. "Stubs" left on the plant will quickly decay and invite fungal attack. Also remove any foliage that is dry or yellow along with insignificant flower buds throughout the bloom phase of development. Be certain to take great care in the trimming and grooming

Cannabis Pruning Primer
less is more

- Use gloves when handling living plant material.

- Use only the sharpest blades when cutting.

- Sterilize shears occasionally or between cuts.

- Use pruning to train and shape plant structure, or cut top to redirect upward growth.

- Make individual, precise cuts; never shear a marijuana plant like a hedge.

- Avoid damage to leaflets.

- Groom plants regularly by removing unwanted foliage.

- Do not prune a stressed plant.

- Water thoroughly before and after any significant pruning.

- Cut no more than ten percent (10%) of a stem or branch at any given time.

- Remove as much of a leaf stem (petiole) as possible when cutting.

- Remove dead or diseased growth immediately.

- Promptly remove any plant structure that has a mite infestation beyond control.

- Remove any plant tissue that shows signs of necrosis.

- The more mature the plant, the less pruning/shaping should occur.

- Discard all pruned material away from plants in cultivation.

process in the latter stages of bloom maturity. There is a tendency to move branches to gain access to plant interiors while clipping, but this procedure requires minimal movement and gentle expertise. It is also easy to poke yourself from sharp pointed shears; do not hurry and make each cut precisely.

When growing cannabis in crop production, give your plants what they require to grow to maturity with a minimal amount of handling. Use the basic, tried, and proven techniques instead of reinventing the wheel of cannabis culture and use your creativity to refine the process. Automate what you can about irrigation and climate control and monitor everything between; all will be the best use of your time in farming for a premium result.

Grooming is Not Heavy Pruning

Removing medium to large sized leaves is a grooming technique in some kinds of cannabis cultivation to allow better light and air penetration to plant interiors, but it is primarily to promote localized growth. This procedure is selective leaf removal of bigger leaves, not all of them; total defoliation will stress and harm the plant. With the continual big leaf removal process, branches and stems swell and bud sites increase. As the plant matures and blooms, the flowers (buds) are more resinous, denser with greater weight and size when the plant has less leaves from continuous removal. For a premium cannabis grower, it is the objective for best commercial value.

The practice is about weekly when plants are young, but as they mature, the leaves increase in proportion and the task then becomes a daily procedure to keep up with it. This is when labor demand increases as well. It is prudent to also remove unwanted growth when grooming, especially from the space near the trunk in the plant interior where small buds will not mature with significance. Spindly growth and small stem removal from the inside of the plant structure is a light form of pruning, but in all the clipping, do not confuse grooming with substantial pruning; one is beneficial while the other is potentially harmful.

In this context, pruning a plant consists of removing branches or tips to train or shape; a practice that is very tricky on cannabis for a couple of reasons. Removing any plant growth at the branch tips may have a result that you may not want. It stunts the plant for a short time but also decreases the likelihood that the branch will develop into a lengthy cola, a trade term for long branches encrusted with dense buds. Instead, although it may have many flower buds, it will usually remain shorter or stubby lacking enough vegetation time to lengthen with additional branching and bud sites. The plant may also send branching from the cut point into a direction that you do not want, like to the inside of the plant, or it may not respond at all. So, unlike perennial plants or trees where pruning is to shape and promote controlled growth, cannabis reacts unpredictably to radical cuts to the structure, so as a rule, keep pruning to a minimum for this short cycle crop.

There are always exceptions to growing rules in cannabis and one here is regarding topping; a procedure where pruning the top leader terminates an upward growth pattern and directs the growth to the remaining plant and branch system. There are many theories about when to top a plant; the best time is prior to the end of the vegetative phase of growth. With ceiling limitations in an indoor cultivation project or in a greenhouse, topping may be necessary to reduce plant height, but it often yields more substantial bud formations and is a common practice for premium flower growers.

Best results in grooming come from cutting the petiole (leaf stem or leafstalk) as close to the branch as possible, without damaging any adjacent plant structure. Cut, and never tear or pull to remove leaves. Sharp, needle point shears are useful for this procedure, but they can easily prick a main branch and cause damage, so use care. In grooming or pruning, it is imperative to remove every scrap of material that you cut. Stems and fragments of leaves are ripe for a fungal disaster for your garden if left on or near your plants, especially for greenhouse or inside cultivation, particularly in the bloom phase of development.

Chapter 4
Media: Soils, Substrates, and Amendments

Hard work and expense go into the cultivation of marijuana, so you want no compromise in the material that will support your plants throughout their lifespan. Aside from your work tending the plants, and the cost of equipment and supplies like fertilizers, it is the primary material investment for the life quality of your plants, from seedlings to mature beauties. Do not cut costs when selecting soil and amendments and you will see the rewards at harvest time.

Soil or the substrate that you choose for your system plays a critical role in providing the support and delivering the other essentials for growth. It is the starting point for a bountiful crop based on your choice of growing method. In choosing a good quality product, cost is not the only determining factor. Also consider what will be the best growing media for your specific cultivation technique and it will serve your plants well.

Once you have invested in your growing medium, keep it alive and vital with a balance of nutrients and maintain optimum conditions for healthy plants and premium yields. This Chapter helps guide you through the wide array of products to navigate selection and maintenance of a desirable growing medium for most cannabis cultivation projects.

4 | a. The Ultimate Growing Media
Choose the Best Based on Cultivation Technique

It is challenging to cultivate premium grade marijuana unless the growing medium (a.k.a. substrate, soil, soilless mix) properly supports the plant and is conducive to delivering water and nutrients at the correct levels for optimal vigor and health. Use what is most productive for your type of cultivation method, despite cost, or convenience, or your results will likely be disappointing.

The selection process for the soil and grow mediums best suited for cannabis cultivation should be methodical if you want the most ideal composition for a premium crop. The decisions you make before you plant a seed or clone are very important to the long-term success of your plants according to the type of farming practice you will use to cultivate your crop.

The method you use to cultivate your plants is the most significant factor in selection of the growing medium. There are two basic types of growing substrates in cannabis horticulture.

•Soil: Refers to field soil. (In the context of cannabis cultivation, "soil" is the material that occurs naturally, outdoors, although some product labels say "Potting Soil." These are mixtures that rarely contain any field soil and are usually mixed.)

•Soilless: Manufactured mixtures of organic or inorganic materials. Use in containers and hydroponics.

LOCATION

Where your plants grow determines the formulations, or mixes, of the soils and amendments.

•Outdoor

-In the ground

Plant directly in amended planting holes or use planting bags containing potting soil designed for placement into the ground at varying depths.

-In containers or fabric pots

Use prepared potting mixes as-is, amended, or custom mixes purchased in bulk.

•Indoor

(A *climate controlled, inside growing space, including greenhouses.*) Choose soilless mixes suggested for indoor use and your cultivation method. Custom mixes suitable for specific cultivation techniques might include the addition of washed gravel, clay pebbles, perlite, or vermiculite to pre-made soil products, while the use of these hard substrates may also be alone. Coco coir is an increasingly common supplement to custom mixes and regularly used as a primary component.

Note: Do not bring native soil from outside to an interior growing space and only use products suited for indoor use when growing indoors, in a greenhouse or hydroponically. Protect soil products of any type from weather extremes and pest contamination.

ORGANIC OR SYNTHETIC

In cannabis cultivation, you have the option of growing your plants organically (without artificial chemicals added) or by using products that contain synthetic formulations to enhance and control plant growth. While it is possible to use both organic and synthetic products together, the practice is risky since over-fertilization is a leading cause of plant damage and failure. If you plan to label your product as organically grown, you should not include substances not listed as organic, even in small amounts.

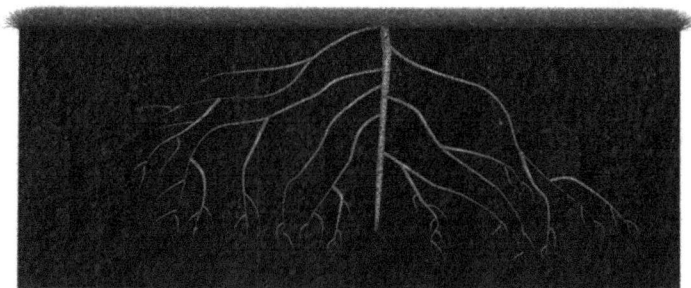

The Perfect Substrate

One of the most common questions at any sales counter where cannabis farmers might trade is "what is the best soil?" The answer is usually a series of questions, like; how are you growing, where and what strain?

Not all soil, or "dirt" is equal and in the realm of creative combinations in soilless mixes, the perfect medium to grow awesome cannabis is what works for your method of cultivation. Never rely on one or two opinions, because most soilless media products on the market today will grow cannabis. It may require amending or correction in some manner, but cannabis is very forgiving and will survive, even in poor soil.

If you are a grower who wants top-shelf bud, then the selection of soils and soilless products narrows to a shorter list of suitable substrates. Where you can purchase the products also is more restricted since the better media and amendments cost a bit more than those sold at garden discounters and are available at specialty nurseries and hydroponic stores. The cost and inconvenience, if any, are worth it considering the likely guidance you will receive from the store you do business with. More important, the biggest benefit is in crop yield; bigger, heavier, and frostier buds result from cannabis that grows in the substrate or amended soil that is applicable for the culture method and location.

If you require more than a reasonable number of bagged products, reach out to your local soil, and landscape supply companies and inquire about custom mixes. Some soilless media companies also sell in bulk, although most have minimum purchase amounts, and you may need a forklift for pallet work. The cost savings when buying in bulk are substantial even with delivery charges and you can even develop your own blend. If you use one-hundred-gallon fabric containers; for example, which may require half a cubic yard of soil per pot, it is really the only way to go. Also, if you are amending for field planting or individual hole planting, bulk purchase of amendments is also an economical and labor-saving solution.

The ultimate definition of the perfect substrate for cannabis is what produces the best crop. Given all the variables, refine your substrate, despite your cultivation method, each time you start a new cycle of production plants.

To make an informed decision about what method of cultivation you will utilize, consider your production goals, your budget, and your facilities. Ask for advice from friends, colleagues, local dispensaries, and sales people at your favorite suppliers. Try to identify trends in consumption that coincide with your own production standards and criteria. Think about the chemicals you use on your plants as they grow and then later ingested by human consumption. As medical cannabis patients look for responsible and accurate labeling, many dispensaries and retailers test for impurities present in the plant material and reject any product that they deem unsuited for their patients, an important consideration if you plan to become a supplier. Also think about the trend in extraction popularity for use in other products, where testing is a part of the process.

If it is not already a requirement in your location where medical or recreational pot is legally available, when contaminant testing becomes mandatory, products for sale must include a label with the testing results.

Whatever method you decide upon, organic, with synthetics or a combination of both, always read and follow the label instructions of the products you use; conserve whenever possible; recycle and more than anything else, realize you are a temporary steward of the land you work on and it requires your respect to receive any reward from it. Not "New Age"; just common sense.

OFF THE SHELF OR CUSTOM MIX

•Ready-Made

The marketplace offers a wide selection of premium soils and amendments from which to choose. Many growers use products especially suited for cannabis cultivation and readily available from hydroponic garden stores or nurseries. Others use products from a variety of sources and mix them for their own magical blend according to recipes they have developed, found on the Internet or at cannabis events. Base your plan primarily on local availability since shipping soil and amendments can be costly because they are typically bulky and heavy.

Although retailers with old stock would disagree, there is a viable shelf life to soils and most amendments so it is good to purchase only what you will need in a season or growing cycle. Volume buying to get a discount rarely pays off if you plan to store your materials for longer than six (6) months, or from growing season to the next. Problems from decomposition and pest infestation are the chief reason for avoiding the storage of your soil products, not to mention the need to protect them from the elements, especially from water and wide temperature changes. Usually, your favorite local supplier can get what you want most times of the year, so there is no real need to "stock-up" on soil products, except for convenience.

•Design Your Own

Every growing situation is different, requiring modification and adaptation of the soil or growing medium to fit the needs of the plants in culture. What works for one grower might not be great for one in another part of the State; for example. Even within the same county or parish, water, and microclimates will affect soil, or soilless mediums differently. Specific cannabis strains also have varied soil preferences.

In the pursuit of excellence, growers are always looking for ways to improve yield quantity and quality. The soil or soilless medium that performs best for your setup and strain selection may be what you create yourself. Water holding capacity in soil is vital for plant survival, so prioritize this aspect. Strive for good texture, correct pH, and nutrient content. Beyond those basic criteria, preferred soils can effectively remove impurities, eradicate disease agents, degrade contaminates and convert dead organic matter into nutrient forms needed for plant growth.

There is a true classic in the world of cannabis and that is the famous and widely used organic soil recipe developed by TGA™ Seed Company. It appears in publications and on the Internet as a mix designed for vigorous growth. It requires at least thirty (30) days of composting once it is thoroughly mixed, but the resulting soil significantly reduces the need for additional fertilization during the life of the plants.

Near the end of the book is a directory of soil and soil amendment companies that make products useful to the cannabis farmer. The ultimate grow medium for any marijuana plant is simply what ultimately produces the best crop. Whether that comes from packaged merchandise, a modification of the ready-to-use products, or a completely unique mixture made from individual ingredients, great grow mediums come from great growers who understand the importance of nurturing their crops by working with their soil and keeping the habitat alive.

Conducting your own research and development program can start with substrate experimentation. Each time you begin a crop, plant a few that are "extras" or not critical to production numbers, in growing media that you have never used before. That can be from a commercial product already mixed or one that you create yourself. The results will likely amaze you and have an influence on future media selection; you will either keep and not change your present substrate or you will improve it using your discovery. In both cases, you are testing for improvement; a strategy essential to long-term success.

There are areas in marijuana farming where an economical approach is prudent; however, when it comes to the growing substrate, spare no expenses. It is the one cultivation commodity that really should be the best quality you can afford, ideally suited for growing premium quality cannabis using your specific cultivation method.

W e have all heard the expression; "You get what you pay for." This is a sound consideration in cannabis cultivation where every expense in the growing process should produce desirable results from the best values available in the supplies you use to grow your crop. Considering the soil or medium used on your farm as an asset or "capital" in your cultivation projects is the first step in comprehending just how important it is to your success.

ECONOMICS OF GROW MEDIA

When you budget for soil or growing media, it is a purchase that will be of multiple units to start and successfully grow a specific number of plants. The initial expense might appear as something that is reducible after adding it up, but soil or your growing medium will be the least expensive part of your cultivation project. Just like any good investment, you want to get the most for your money when buying soil, growing medium, or amendments. Do not assume that the most expensive is the best for cannabis cultivation or that correction of poor soil quality is possible with fertilizers and additives. If you are looking for top-shelf bud, what your plants grow in is critical to success and underestimating the need for healthy soil will lead to unhealthy plants and disappointing crop yields.

Like money in an investment that you would monitor from time to time, after your purchase you want to make sure to regularly tend to the soil or medium so that you provide the optimal conditions for your plants to grow well in the vegetative cycle and achieve maximum yield in the bloom cycle. From how you store it, use it during your grow or recycle it after harvest, it is important to have a plan for the soils, mediums, and amendments for the entire life cycle of the cannabis plant. The key in your plan is the nutritional program you establish because despite the soil or medium you select, it will require regular monitoring for good health and a replenishment of nutrients necessary for the best results in growing cannabis. Starting off right, at the ground level so to speak, and then maintaining the soil properly is the best way to attain a premium product and receive a good return for your expenditures of money and work.

There is no perfect soil or medium for cannabis and if anyone tells you that there is, run. There are so many variables in growing techniques and conditions, not to mention varying requirements by different cannabis strains, to call any soil, medium, amendment, or even a soil recipe, perfection is a misnomer. The general characteristics of soil or growing medium required for optimal growth is good draining but with the ability to hold water well, nutrient rich and a pH range of 5.8-6.5. How you achieve the specific soil conditions ideal for healthy, vigorous growth are the secret to success in cultivating high-grade cannabis and the key subject of this book.

To better choose a soil or amendment program and then to properly manage your soil or soilless medium requires a basic understanding of what role growing media play in your pursuit of premium grade cannabis. It is a big deal because soil or grow mediums perform functions that are essential for plant growth. They include physical support for the plant, supply air, and water, provide temperature modification, nutrient supplies, and protection from toxins. Beyond these basics, good soil and proper management of that soil play important roles in the composition of the product; bud density, color, texture, flavor, aroma, and THC content are all influenced by the growing medium.

COST TO BENEFIT RATIOS

Commercial growers consider cost to benefit ratios when developing a purchase plan and the principles are good to follow for any size grow. Given the increased value of a crop using good management practices, soil is a small percentage of the actual growing costs described in the Retail Costs analysis, below.

A good quality product does not need to be expensive, and you should not equate price with suitability. When shopping for a soil or soilless medium, look first for the components that will be compatible with your growing system and consider your production goals.

When making your final purchase, look for the best price. Ask for discounts on volume purchases and negotiate delivery charges if possible. If you have an established relationship with a supplier, chances are you will get a good deal. Whatever the price, know that it is a sound investment in your cultivation project.

It is not a myth that good farmers have a partnership with the soil that produces their crops. By starting with the best grow medium that you can provide, the symbiotic relationship between the soil, the plant, and you will pay off handsomely when it is time to harvest your premium crop and you realize the cost of your soil or soilless medium was worth every dollar.

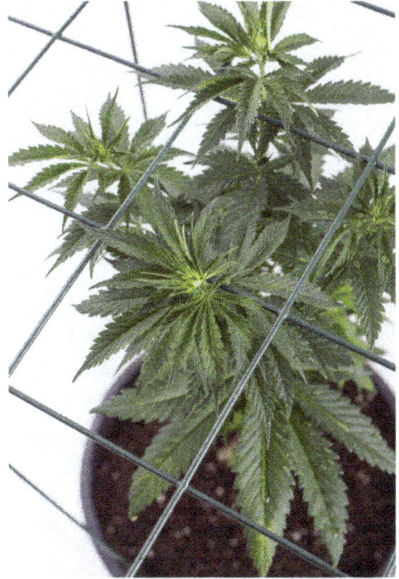

Retail Costs of Basic Start Components and Final Crop Value

Prices are average costs at publish time; hypothetical and may not apply to your locale.

Merchandise

a. Clone in two-inch rock wool: $14.00 (33%)

b. 10-gallon container: $6.00 (14%)

c. Starter Fertilizer: $1.29 (3%)

d. Bamboo Stakes: $1.00 (2%)

e. Soilless potting mix, 1.75 cu. ft.: $19.96 (47%)

   Total Start Cost for one (1) unit: **$42.25**

Crop Value

   Eight (8) ounces of flowers, trimmed: **$1,920.00**

   Cost of soilless mix to produce eight (8) ounces: **$19.96 (1%)**

Looking at the cost of a good quality growing medium, it is almost half of the cost of setting up one unit, or container in the breakdown above. However, when compared to what the unit will produce, it is a very small portion.

4 | c. Nourish and Amend for Fertility
Optimization for Every Growing Medium

Tending the soil is a continual process required to grow premium cannabis crops. Experienced farmers know their plants are removing nutrients and restoring them in some manner is necessary for the substrate to remain viable for maximum crop production. Even super soil recipes will require attention to keep minerals in balance to prevent over fertilization or create nutrient deficiencies.

In a broad definition, soil fertility is the ability of a soil (or growing medium) to supply nutrients to plants that are essential for growth. In a narrower explanation for cannabis horticulture, fertility is the value of flowers (buds) produced relative to the cost of the amendments used. In simple terms, a cannabis grower considers fertility of a soil as its capacity to produce a good quality product; in weight, potency (THC and CBD), flavor, and aroma (nose).

Soil fertility is measurable, but also subjective. For example, food crop farmers who utilize the same fields often consider sustainability in assessing the fertility of soil, while cultivators of nursery stock in containers might consider the humus content as a chief indicator in gauging fertility of potting soils, etc. It is important to understand that aspects of fertility are usually contextual, but in all cases, fertility is dependent on physical, biological, and chemical characteristics, and the interactions of these three parts. For best results in cannabis cultivation, fertility goals include a soil that has a good water holding capacity with ample air spaces, good drainage, supply the correct levels of nutrients and it should team with beneficial organisms.

Rarely is any soil optimal for specific use or crop production without some modification or amendment. Since the nutritional requirements for plants have so many variables, identify the needs of individual crops, or in cannabis, the strain, and then develop a soil management plan adopted to meet those needs. Growers use several methods with various materials to improve soil for general crop production.

These techniques/materials are useful in fertility adjustment and soil conservation:

- •Crop rotation
- •Cover crops
- •Composts
- •Fertilizers
- •Micronutrients
- •Soil Conditioners
- •pH Adjusters
- •Beneficial Organisms

SOIL MANAGEMENT PLAN

Listed are goals and procedures that should be in a written plan to manage the soil or growing media in cannabis cultivation. The advantages of having a concrete plan will not only make operational management easier and efficient, but growers can enjoy the economic benefit from the better crop yields, as well.

Objectives: Protection of the soil/media and enhancement of it for the benefit of both plants and soil.

Practices:

- •Tilling/aeration (outdoor) when erosion from weather is minimal
- •Use of cover crops (outdoor)

•Rotation of crop location (outdoor)

•Amendment - Develop plan before planting

•Irrigation - Develop a watering schedule that is efficient with minimal waste

•Fertilization - Use a feeding schedule through growth and bloom phases

•Pathogen and Pest Control - Schedule regular exams of soil and plants

•Disposal - A safe and responsible method of discarding spent media

AMENDMENTS USEFUL IN CANNABIS FARMING

Soil amendments, or conditioners, are materials added to the soil or growing media and should serve to improve the soil structure, water retention, drainage, and aeration. Amending soil is the addition of material, mixed into the soil, but do not confuse it with mulching, which is typically the addition of material on the surface of soil, to retain moisture, and suppress weeds. Growers may select from both organic materials (natural) or inorganic (processed or manufactured).

When selecting an amendment for use, consider how long the amendment will last in the soil when mixed (mix all amendments thoroughly with soil that will be in the root zone), the pH measurement, salt content and cost. Since bagged or bulk products for use in soil are usually unregulated in most areas, use plant based composts (low in salt) when possible. Use caution with animal manures and do not use any manures not aged or processed; they may be caustic to your plants.

Some amendments, especially from animal origin, emit odors as they decompose in the soil or medium, and may not be suitable for indoor use.

COMMON AMENDMENTS IN CANNABIS CULTIVATION

• **Plant Origin**
Alfalfa Meal
Biochar
Coco Coir
Coffee Grounds
Compost
Cottonseed Meal
Peat (Sphagnum moss)
Straw
Wood chips

• **Animal Origin**
Bat Guano
Bird Guano
Blood Meal
Bone Meal
Earthworm Castings
Fish Meal
Manure

• **Manufactured/Processed**
Azomite®
Greensand
Perlite
Polymers (Hydro-absorbent)
Sulfur
Vermiculite

FERTILIZATION

Provided nutrients are only useful when minerals and compounds in a solution or substrate are working together in the right amounts, available and applicable for the varying stages of plant growth. With an understanding that effective fertilization is nourishment best suited for the cultivation method, works in harmony with the substrate, and changes by growth phase, the easier your path to a good nutrient program and an awesome crop.

Fertilization of cannabis is a complex subject. There is an immense number of variables in the cultivation of any plant, and marijuana is no exception. As one might expect, there are just as many fertilizer products available in the marketplace as there are variables in cultivation. For new growers, the temptation to rely on colorful packaging, wild marketing claims or the opinion of someone on the Internet for product selection is very common. Even for experienced growers, new products are of interest as most cannabis cultivators are continually looking for crop improvement.

A solid piece of advice; get a basic understanding of fertilizers and growth enhancers, then choose a product line, and stick with the array of products in that line. There is a wealth of information to help you on various manufacturer's web sites and spending just a few minutes on those who make products you are considering is time well spent. Companies develop their fertilizers and additives to work in harmony with one another and to be compatible through the vegetative, transitional, and bloom phases of growth. The reputable manufacturers produce science based products you can rely on.

Fertilizers are available in liquid form for dilution, or in dry form for gradual release after application. Classification of fertilizers is in several ways:

•**Straight Fertilizers**: Provide a single nutrient, usually to correct a deficiency.

•**Multi-Nutrient Fertilizers**: Provide two (2) or more nutrients. The most common is formulations that contain three (3); N-P-K.

•**Inorganic (Synthetic) Fertilizers**: Do not have carbon containing compounds. (Except Urea).

•**Organic Fertilizers**: Made from animal or plant materials, without chemical or artificial treatments in the manufacturing process.

Multi-Part Fertilizers

As nutrient products have become very sophisticated and precise in their function, some manufacturers produce two (2) or three (3) part liquid formulations. These fertilizers begin a chemical reaction when mixed in water prior to application, and then used immediately. Some contain coloring tracers for monitoring. These fertilizer systems are common in many indoor and commercial cultivations. Most producers of multi-part fertilizer systems also offer feeding schedules with exact dilution rates according to the stage of plant growth. These are highly reliable and offer practical guides to proper fertilization based on the most current crop science.

FUNCTIONS OF MINERALS IN FERTILIZERS

•Macro nutrients

Nitrogen (N)

Part of chlorophyll, amino acids, and proteins; helps plants with growth; increases seed and flower production; improves leaf and tissue quality.

Phosphorous (P)

An essential part of photosynthesis and energy transport; assists in formation of oils, sugars, and starches; helps withstand stress; encourages blooming and root growth.

Potassium (K)

Absorbed by plants in large amounts; helps build protein and assists in photosynthesis; helps in bloom quality; helps resist pathogens and disease.

Calcium (Ca)

Essential part of cell wall structure; helps in transportation of nutrients; adds plant strength; helps to regulate enzymes.

Magnesium (Mg)

A part of chlorophyll; required for photosynthesis; helps activate plant enzymes.

Sulfur (S)

Assists chlorophyll production; helps resist stress from cold; helps in chlorophyll formation; essential for protein production; improves root growth.

•Micronutrients

Boron (B)

Assists in sugar and carbohydrate production; required for seed and flower development; assists in nutrient regulation like Calcium (Ca).

Copper (Cu)

Assists in root development; protein uses and reproductive growth.

Chloride (Cl)

Assists in plant growth; ionic balance.

Iron (Fe)

Required for chlorophyll production, effective in treating chlorosis.

Manganese (Mn)

Breaks down carbohydrates; assists in enzyme functions.

Molybdenum (Mo)

Assists in nitrogen use (fixation).

Nickel (Ni)

Assists in enzyme functions; nitrogen metabolism.

Zinc (Zn)

Helps to transform carbohydrates; regulates sugar consumption; assists in enzyme functions.

NUTRITIONAL SUPPLEMENTS

With a goal of high yields and a quality product, cannabis cultivators have a wide range of products to balance and enhance plant growth, besides a regular fertilizer regimen for growth and bloom.

The accelerated growth by most cultivation methods and the controlled environments in which the plants grow will influence amino acids, hormones, and vitamins naturally produced in cannabis plants. Nutritional supplements are a way to balance the biological processes to achieve optimum plant health and vigor, resulting in better yields. Additionally, growers can influence the density, flavor, and aroma of the finished bud with products compatible with basic nutrient programs.

Compost Teas

For use in soil, soilless and hydroponic growing, compost teas brew with humic acids to create unique formulations that effect flavors and aromas of cannabis bud; readily available and popular for organic growers. Like other organic fertilizers, compost teas may attract wildlife like raccoons. When used indoors, some organic nutrient products may emit an odor and may leave a residue on irrigation components.

Calcium-Magnesium; a.k.a., "Cal-Mag"

A widely used supplement for flowering plants like cannabis, Calcium-Magnesium compounds are excellent for compensating for the absorption of Calcium in coco coir growing media. Note that calcium and magnesium are the two most common minerals that make water "hard".

Bloom Stimulator

The bloom phase of cannabis growth demands more than most N-P-K formulations can offer for best results. Specialized supplements are available that contribute to larger and more dense bud formations, largely through stimulation of vitamins and enzymes.

CannaBananaSuperGro
10 - 4 - 8

GUARANTEED ANALYSIS

Total Nitrogen (N)...	10.0%
6.56% Ammoniacal Nitrogen	
3.44% Urea Nitrogen*	
Available Phosphate (P_2O_5)...................................	4.0%
Soluble Potash (K_2O)...	8.0%
Total Magnesium (Mg)...	1.2%
1.2% Water Soluable Magnesium (Mg)	
Sulfur (S)...	7.0%
7.0% Combined Sulfur (S)	
Boron (B)..	0.02%
Total Copper (Cu)...	0.05%
0.01% Water Soluble Copper (Cu)	
Total Iron (Fe)..	1.5%
0.01 Water Soluble Iron (Fe)	
Total Manganese (Mn)...	1.4%
0.01% Water Soluble Manganese (Mn)	
Molybdenum (Mo)...	0.0005%
Total Zinc (Zn)..	0.05%
0.01% Water Soluble Zinc (Zn)	

Derived from: Polymer-coated Urea, Urea, Sulfate of Potash Magnesia, Muriate of Potash, Sodium Borate, Copper Sulphate, Copper Oxide, Ferrous Sulfate, Ferric Oxide, Manganese Sulfate, Manganese Oxide, Zinc Sulphate and Zinc Oxide.

*Contains 3.1% slowly available nitrogen from coated urea.

Directions for use:..
....
....

CannaBanana Farm Fertilizer
Cloud 9 Avenue
Any Town, Any State   **NET WEIGHT 25lbs.**

This fictitious fertilizer label shows what federal law requires; Brand name, Grade Statement, Guaranteed Analysis, Derivation Statement, Directions for Use, Name and mailing address of registrant, distributor, or manufacturer, Net Weight, or Volume. (*The company and formulation do not exist and not intended to resemble any existing entity or product.*)

Growth Enhancers

Specialized formulations contain organic compounds not found in other fertilizers by utilizing high metabolic activity and quick absorption.

Silica Products

Favored by cultivators growing in coco coir, silica or potassium silica formulations assist in strengthening plant tissues. Products in this category are helpful to plants in handling temperature extremes (hot or cold) or stress from dry conditions. Dry weights of cannabis buds increase with silica formula use as they increase mass during growth and flower formation.

Flavor Enhancers

Formulations are available in fruit flavors that utilize carbohydrates, amino acids, and vitamins to change the taste and aroma of finished bud. Most also contain organic compounds that are beneficial to plant growth. Old-time growers use molasses for "sweetening," although Blackstrap molasses has other benefits including calcium, magnesium, potassium, and iron content for nutrient supply. Molasses also encourages the growth of beneficial microorganisms. Synthetic flavors like grape, citrus, and berry, are available.

Chelators

Helpful in optimizing crop yields, chelator formulations help to attract and hold nutrients at sites where the intake is most beneficial. Chelators are also useful in protecting plants from stress. Some are useful in nutrient solutions while others are suitable as foliage sprays.

The concept of optimization for a better crop is simple. If you begin with a good soil or growing medium, the task of enhancing the root ecosystem for your plants becomes one of a stimulated balance. It is "age appropriate" in that the balancing act changes as the plant matures from a seedling or cutting to a blooming beauty. Plan with a nutrient spec sheet that details each week of the grow cycle and have the products to enhance your medium on hand before you need them.

The wide assortment of soils, substrates and amendments in the marketplace make for numerous opportunities to provide the ideal growing medium for every cannabis cultivation method. Choosing the right products and combination of components are the key to delivering what the plant needs for healthy growth and abundant production.

A wide selection of soilless substrates used in horticulture affords the cannabis grower several options to achieve healthy, vigorous plants along with desirable production results. Those most commonly used in cannabis cultivation include both natural and synthetic materials or combinations of the materials listed below. Each substrate, used alone, or in a mix has its advantages and disadvantages to consider along with cost and availability, before selection, and use.

Several premium-grade soilless mixes are commercially available. Professional growers often choose these premixed substrates for not only convenience, but because they produce desirable results. Outlined are stand-alone substrates and some of the base components of these mixes.

•Coconut Coir

This natural fiber is from the husk of a ripe coconut (brown) after washing and composting for horticultural use. Coconut coir is an excellent growing medium, chiefly because it has a high water holding capacity and is slow to shrink after saturation. Additionally, coco fibers retain 25-30% of oxygen at the root zone. With a high lignin content, it is also long-lasting, and reusable. Coco coir does not degrade, maintains its structure and is extremely versatile.

-Pro

Good water capacity, good rates of air porosity, high cation exchange capacity.

-Con

Product requires charging to adjust for high levels of potassium before use. (Usually completed prior to packaging or sale in horticultural applications.) Nutrient solutions usually require some adjustment; especially the Calcium and Magnesium levels.

•Growstones®

An alternative to expanded clay aggregate, Growstones® are from recycled glass. Their irregular surface areas and porous structure provide more air and water retention space than either peat moss or perlite. This product is sustainable with several technical bulletins available on its use from University trials. This is a widely used product by cannabis growers; consult seller for product availability and size options.

•Hydroton®

This is a manufactured aggregate from expanding clay in a rotary kiln at high temperatures. The gases expanded in the process leave voids that are helpful in retaining water and air spaces for roots. The round shape formed when fired in a rotary kiln adds additional space when used in potting containers, alone or mixed with other media. (Note: Hydroton® clay is a reclaimed material from building waste. Although widely used, they are not specifically for horticulture.)

-Pro

Good air porosity, reusable.

-Con

Inert with no cation exchange capacity, dust removal required prior to use, by rinsing.

•Hydrocorn®

A trade name product certified for horticultural uses. Like Hydroton® in appearance and structure, Hydrocorn® is from pure clay and fired in an open furnace using clean burning fuels. Beneficial bacteria and fungi thrive on a clay pebble environment, making their use attractive to cannabis growers. Like Hydroton®, Hydrocorn® requires rinsing of clay pebbles before use.

(Hydrocorn® holds certification by the RHP foundation for horticultural use in the Netherlands.)

-Pro

Good air porosity, reusable, certified for food crops in horticulture.

-Con

Inert with no cation exchange capacity, may cost more than Hydroton®.

•Peat Moss

Sphagnum moss decayed and dried has the name; Peat Moss. Widely used for horticultural purposes because peat can hold water and nutrients by increasing cation exchange capacity (CEC) and capillary forces. Peat moss is acidic, a quality that helps inhibit harmful bacterial and fungi growth; ideal for shipping live plants if used alone. Peat may or may not be a renewable resource as worldwide data suggest. Coir is fast becoming a sustainable alternative to Peat Moss in horticultural use for specialties such as cannabis.

-Pro

Readily available at most garden stores, good water holding capacity with good air porosity.

-Con

Poor aeration, not sterilized, correct pH prior to use for growing cannabis.

•Perlite

A heated volcanic glass produces an industrial mineral with numerous uses, including filtration. In horticulture, perlite is common as a soil amendment to improve water and air capacity, or alone for propagating clones (cuttings) or for use in hydroponic growing. Although it is a lightweight and inexpensive, it is a non-renewable resource.

-Pro

High porosity with high water holding capacity and excellent aeration, providing air to roots.

-Con

May clog drains or pumps in hydroponics, low cation exchange capacity and cannot hold nutrients in reserve.

•Rice Hulls

The outer part of a rice seed is hard materials like silica and lignin. In cannabis cultivation, rice hulls parboiled are an amendment to improve soil tilth and drainage without affecting plant growth regulation. Although slow to decay or compost, rice hulls are not a reliable stand-alone substrate in growing cannabis.

-Pro

Biodegradable; readily available. High air porosity ratio.

-Con

Poor stand-alone substrate, low cation exchange capacity.

•Rock wool

Also known as mineral wool (also spelled rockwool) rock wool is a manufactured fiber produced from spinning molten minerals. Rock wool holds large amounts of water and air making it most useful in cannabis cultivation using hydroponic methods. The structure of rock wool provides a good support mechanism for plant stability, and the fibrous composition is excellent for aiding in root growth and nutrient uptake. Rock wool has a higher than optimal pH for cannabis and requires conditioning before use.

-Pro

Free from plant pathogens and provides high air porosity ratios.

-Con

pH requires adjustment, non-reusable and disposed after use.

•Vermiculite

A hydrous mineral, vermiculite comes from mines. When heated, vermiculite significantly expands and exfoliates to produce a material with wide commercial uses. In horticulture, vermiculite combines with other materials to produce soilless mixes, or as an amendment to improve drainage and air capacity in clay or sticky soils. Some cannabis growers utilize vermiculite for seed germination with ultra-mild fertilizer applications.

-Pro

Lightweight, readily available. Provides good drainage and lightens heavy soil.

-Con

Low water holding capacity and low cation exchange capacity.

PURSUE THE IDEAL MIX

Fortunately for the professional marijuana farmer, the wide variety of soilless growing mediums has a product for every need and system. Various mixtures are for use as a growing medium in containers, as a soil conditioner for other media and for amending soil for in-ground planting.

Regardless of how you use them, most soilless mixes are customizable to help retain air, plant nutrients, and moisture, releasing them as the plant requires them. After starting with good base mixes, finding the right combination for maximum benefit is the key to success for your individual farming situation. Like many aspects of cannabis cultivation, there are always ways to improve and experimenting with the materials available to you will often lead to you to better products for your plants resulting in better crops at harvest time.

Coir

The beneficial uses of coconut fiber are numerous. For centuries, this natural fiber was the source of ropes, cords, and strings. A natural product from coconut husk, brown fiber coir is in many soil mixes that cannabis farmers favor.

After removal from the coconut shell, millers separate the coir after soaking the husks in nets suspended in water. Sorted by fiber size and length, the fibers dry in the sun, with additional rinsing, usually by rainwater, for horticultural use. Coconut coir pith is high in sodium, and potassium compounds so treatment for plant use requires placement in buffering solutions containing calcium. Even so, planting media containing coir remains rich in potassium, but requires the addition of both calcium and magnesium to make it suitable for plants.

The high lignin content of coir makes it desirable for horticultural uses because it is long lasting and holds more water than peat. It is a light weight media, and it does not shrink making wetting the material very easy, unlike peat moss. Environmentally friendly, coir helps prevent over-watering in container mixes; helpful to marijuana plants that have a low tolerance to soggy soil.

There are disadvantages to coir as a substrate, mainly that it is an allergen to some people and may host various species of harmful fungus. The pathogenic fungi may exhibit in greenhouses requiring treatment; however, the addition of beneficial microbes to soilless mixes helps in mitigation of the problem. Coir from Mexico contains beneficial fungi colonies that help detour pathogens to plants, so the source of coir is very important to growers. If the source is unknown, choosing reputable, commercially available coir specifically for plants is prudent for all cannabis growers.

Alone or mixed with other materials like sand and compost, coir is a sustainable product unlike peat that requires mining. If you rely on prepared soilless mixes or use coir in your own custom mix, add the correct compounds to balance nutrient deficiencies and correct the pH for best plant health and benefit.

Good Cannabis Farming Is Sustainable Agriculture

The scope of modern day agriculture is complex, but there is one strategy that is a common denominator in what most smart farmers strive for despite the crop they grow; sustainable agriculture. Cannabis is not a food or animal product like other crops in farming, yet the smart cannabis grower is no less a steward of the land.

With techniques that consider the environment, public health, and the greater good of humanity, the expanding ranks of cannabis farmers create an opportunity to step beside the leaders in ecological food crop production and use their best practices to make sustainability a priority in our operations as well. From a strictly business standpoint, it makes good sense, and from a reference of caring about our colleagues, it is the only route to take.

The benefits are enormous; from our wallets, to the environment, to our local communities and for the credibility of an entire industry. Most will concede that growing marijuana is a significant task on many levels. Taking a deep-seated passion for growing to the next level using responsible practices gives that feeling justice and paves the way to a bright future as we protect our natural resources.

The subject of this book is about ways to be a smart farmer and attain crop goals of quality and integrity, but always with the bigger picture in mind. Smart cannabis farming is sustainable farming and a powerful way to success and achievement in the marijuana industry, no matter what part of the planet you call home.

"We do not inherit the land from our ancestors, we borrow it from our children."
-American Indian Proverb

Chapter 5
Crop Management: Objectives for Sustained Viability

The soils, substrates and environments used for marijuana cultivation that is ideal to flourishing plants, are also host to other biological entities harmful to cannabis. To handle the challenges of farming in a paradise of biology, growers must remain vigilant in prevention, observation, and treatment for the pests and pathogens in their gardens, in ways that are economically sound and ecologically responsible.

An integral part of proper crop management relies on practices that will protect vital resources for future use and not harm the environment when materials are past usefulness for cultivation. In a variety of ways, your farm, or a garden will affect the ecology somewhere, so considering material selection and waste management are sound "green" strategies not only for our global environment, but for your own land and future uses, as well. Responsible objectives and procedures lead to protecting natural resources and sustainability for your water and soil; key goals of conservation and concerning to any serious grower moving forward.

5 | a. Pests and Pathogens
Enemies to A Bountiful Crop

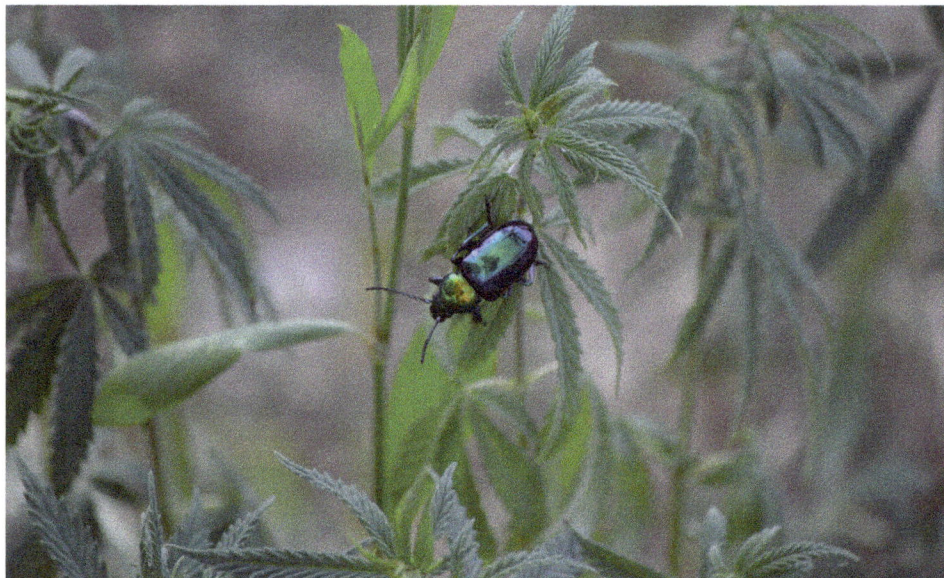

Pest management includes the prevention and treatment of diseases, arthropods and other vermin that pose a risk to the health of your cannabis plants. All can work to cause harm to your crop; individually and as hosts to other pests and diseases. Using a strategic approach to protecting your cultivation investment is a sound practice for any farmer, despite crop, quality, or size.

There is a multitude of pests that can go after your marijuana plants. Early identi-fication is a critical step in control and while not all pests and pathogens appear here, those that are the most common and the most harmful, do.

Misidentification of a pest, disease, or nutrient deficiency is a leading cause of crop damage and failure. Do not rely solely on photographs or descriptions in books or elsewhere to diagnose a problem. Instead, take a sample of damaged plant material, place it in a seal-able container or bag, and take it to a plant nursery, hydroponic supply store, or your local Agriculture Extension office for proper and accurate iden-tification and treatment plan.

BE SMART AND SAFE: Use all pesticides in strict accordance with product labels and consult MSDS for product safety information.

FUNGI

Rhizoctonia solani

Description: Plant pathogenic fungus, soil born.

Distribution: Worldwide; wide range; prefers warm, wet climates.

Symptoms of injury: Attacks hosts in juvenile stages of growth at roots and lower stems causing collar rot, root rot, and damping off.

Controls

•Cultural: Not possible to completely control; select sclerotia free seeds from re-liable sources, choose disease resistant strains, minimize soil compaction, control growing environment.

Pythium

Description: Highly destructive root fungus that multiplies rapidly; control diffi-cult.

Distribution: Pythium exists almost everywhere in a plant environment. Often, termed a secondary infection, it primarily attacks unhealthy plants but will affect and spread to other available hosts.

Symptoms of Injury: Yellowing of foliage and brown leaf edges. Infested plant may appear stunted.

Controls

•Cultural: Maintain an immaculate growing environment, especially in hydroponic systems. Prevent fungus gnat activity; a vector of pythium spores.

•Biological: Use of beneficial fungi (*Gliocladium virens*) (*Trichoderma harzianum*).

Fusarium

Description: Plant pathogenic fungus, soil born. Species varies and can produce different diseases in cannabis, usually fusarium wilt or fusarium root rot. Spores can remain dormant in soil and may disperse by seeds from infected plants.

Distribution: Varies by species; may lie dormant in earlier planted areas.

Symptoms of Injury: (Fusarium wilt) Spotted lower leaves that turn yellow brown, leaves turn upwards. Stems turn yellow, then brown, and eventually crumble or break. Fusarium root rot occurs below soil line with red colored roots that are in a state of decay and necrosis. Disease spreads through plant cells and up the stalk, eventually collapsing the plant.

Controls

Removal and destruction of infected plants are the only effective controls for this disease.

Fusarium infected material should not go in the ground, composted, or placed on uninfected soil areas. Crop rotation is important to prevent fungi from reaching harmful levels in outdoor gardens. Container growing and use of sterile planting mixes is the best ways to avoid this destructive pathogen.

Verticillium Wilt

Description: Plant pathogenic fungus, soil born. Thrives in soils with high clay content or poor draining. Begins by attacking unhealthy root systems.

Distribution: Common; found in moist soils.

Symptoms of Injury: Lower leaves turn yellow along the margins before turning grayish-brown and wilting. Stems turn brown at the soil line.

Controls

Removal and destruction of infected plants or portions of an infected plant are the only effective controls for this disease.

ARACHNIDS

Red spider mite or Two-spotted spider mite; *Tetranychus urticae*

Description: Common plant pest; a plant feeding mite that is dangerous to cannabis.

Distribution: Widespread; it is the most widely known member of the family *Tetranychidae* or spider mites.

Symptoms of Injury: Leaf discoloration, flecking, leaf loss, plant death. Webbing appears after dangerous infestation.

Controls

•Cultural: Mites thrive in arid conditions; use irrigation and moisture management to discourage mites and encourage natural predators like Lady beetles, predatory mites, minute pirate bugs, and predatory thrips.

•Biological: Predatory insects, predatory mites.

•Chemical: Avoid insecticides that may harm natural predators; use miticides approved for crop use; do not use products or horticultural oils approved only for use on ornamentals. Azadirachtin is effective; also sulfur if plants receive thorough dusting at regular intervals.

Broad Mite; *Polyphagotarsonemus latus*

Description: Wide range of hosts in warm regions and a serious pest to agriculture and cannabis plants. Broad mites complete a 4-stage life cycle; 2-3 days as ova (eggs), larva develop into nymphs in 2 to 3 days and become adults in about a week.

Males carry a female larva to new foliage while adults disperse however, possible, including attachment to flying whiteflies. Toxic saliva from feeding mites causes distorted growth and substantial damage often before discovery.

Distribution: Temperate and subtropical regions.

Symptoms of Injury: Malformed terminal buds and stunted growth; often resulting in permanent loss of cannabis flower structures or plant death.

Controls

•Cultural: Minimize plant exposure, quarantine new plant stock. Disinfect growing tools and environments after removing infested plants. Pests may harbor in growing media; discard if infestation occurs.

•Biological: Insecticidal soaps may reduce populations; pyrethrin treatment may be necessary for control. Greenhouses and grow rooms may require air-dispersed insecticide/miticide for eradication. Always follow label instructions carefully.

•Chemical: Miticides suggested for food crops, listing Broad mites as a target pest.

MITE
CONTROL

USE HIGH QUALITY MEDIUM:
Know the source and content of any soil or growing substrate that you purchase. Any amendments must be fresh and free from contaminants including insects.

PRUNE: Remove infested growth or discard entire plant if control is not effective to prevent spread to other plants.

PURCHASE: Secure soils and plants from reputable sources.

QUARANTINE: Keep new plant stock (especially clones) isolated for 10-14 days for inspection before introduction into grow areas.

BENEFICIAL NEMATODES: Add to soil when temperatures warm to destroy eggs and nymphs. A second application may be needed if damage to lower plant is observed.

BIO-SAFE PESTICIDES: Neem Oil will kill and deter mites, while Azamax™ will interrupt life cycle and is useful in bloom stages and near harvest.

DISCARD: Any growing medium that cannot be sanitized must be discarded if mites were present; use local guidelines for disposal.

DISINFECT: Use a dilution of one part bleach to ten parts water (1:10) to clean hard surfaces like containers or pots before use, or use a product called Physan 20™.

INSPECT: Daily examination of new growth is essential for control and eradication to minimize damage to plants.

Hemp or Russet Mite; Family *Eriophyid*

Description: This is a very serious pest to marijuana farmers. The adult requires magnification for visibility; damage often misidentified as a nutritional (mineral) deficiency. This pest multiplies quickly in warm and humid climates, dispersed by the wind in crop intensive locations. Females over winter in twigs and stems and lay eggs in the spring. The nymph stage goes through two phases of development and matures in about 8 days.

Distribution: Temperate agricultural regions of Florida, California, and Oregon.

Symptoms of Injury: Damage occurs near the bottom of plants with yellow or curled leaves and spreads upward. Attracted to trichomes in flowers; this pest will do great damage if left unchecked.

Controls: Prevention is the most effective control of this pest; make certain that plant material from previous grows is not present before planting.

•Cultural: An immaculate grow space and regular examination of plants, especially leaves that appear discolored or malformed can identify infestation early enough to eradicate effectively. Use only dependable growing medium from a trusted manufacturer or source. Newly composted substrates may contain eggs.

•Biological: Beneficial nematodes introduced to warming soils can kill eggs and spider mite predators released indoors or in a greenhouse are effective for early outbreaks.

•Chemical: Neem oil applied at the first sign will repel and kill mites. Azamax® interrupts life cycle and is effective for use near harvest time. Pyrethrum sprays require total coverage for best results, but some compounds derived from chrysanthemums are not suitable for application to cannabis produced for consumption.

INSECTS

Mycetophilidae (Fungus gnats)

Description: A family of small flies with more than 3,000 species. Predatory larvae feed on fungi (sporophores or mycelium). Vectors of plant pathogens. A nuisance in greenhouses and indoor growing areas.

Distribution: Worldwide; damp environments.

Symptoms of Injury: Flies are visible around plants. Nuisance pest; a vector of pathogens.

Controls

•Cultural: Life history from an egg to an adult is 25-30 days, requiring damp organic matter to grow. Reduce these areas to help reduce breeding sites. Allow soil to dry between watering to reduce fungus gnat populations. Screen window and vent openings with fine mesh fabric.

•Biological: Use sticky traps to reduce populations of breeding adults. Diatomaceous earth mixed in soil and applied to media surface around plant base is an effective control. Predatory nematodes are also useful in controlling this pest.

•Chemical: Bt (*Bacillus thuringiensis*, subspecies *israelensis*) safe for use on cannabis plants in vegetative stage. Always follow label instructions carefully.

Family *Formicidae* (Ants)

Description: Most widespread of insect species. Rarely harmful to cannabis; rather, they are symptomatic of other pests like mealybugs or aphids that ants tend if they are present on marijuana.

Distribution: Worldwide; may nest near damp or irrigated areas.

Symptoms of injury: Rarely any substantial damage, but notched leaves indicate leaf-cutting ants are present. More of a problem are aphids that ants may tend for honeydew as a food source. Look for damage by secondary pests (aphids) with evidence like curled leaves or black-sooty appearing foliage.

Controls

•Cultural: Cultivate regularly to disturb colonies but avoid damage to roots.

•Biological: Use pest barriers, diatomaceous earth, iron phosphate with spinosad, or products specific to ant control. Horticultural oils or organic sprays may be useful in discouraging ant populations.

•Chemical: Avoid harsh chemicals around your cannabis plants to control ants. Ants are challenging and usually controlled with agents to protect structures, or non-food crops (ornamental stock). Unless they are causing specific harm to your plants, use biological controls instead.

Mealybugs; Family *Pseudococcidae*

Description: A family of unarmored scale insects with an incomplete metamorphosis. Vectors of disease. Larva and female adults (appear as nymphs) feed on plant juices. Males appear wasp-like, do not feed and are short lived. Serious pests in the presence of ants that protect them from predators and parasites.

Distribution: Widespread; species varies by region.

Symptoms of Injury: Large populations will induce leaf loss and stunt growth.

Controls

•Cultural: Minimize plant exposure, quarantine new plant stock.

•Biological: Ladybird beetle larvae and adults, diatomaceous earth applied to soil surface around base of plants to debilitate tending ants. Some success with using the fungal control agent, *Verticillium lecanii*.

•Chemical: Organophosphates may be effective if approved for food crops but may cause harm to beneficial insect predators.

Root weevil; Family *Curculionidae*

Description: Most damage from root weevils occurs in potted plants or nursery stock by larvae that feed on roots, causing serious damage to seedlings and young transplants, while weakening mature plants. Adults are flightless and feed on leaves, causing a notched appearance. Larval development and pupation occur in the soil. Reproduction is asexual; females lay fertile eggs without mating.

Distribution: Widespread; species varies by region.

Symptoms of Injury: Stunted growth; leaves curl upward. Root system damage allows pulling from the soil with little effort.

Controls

•Cultural: Crop rotation, deep plowing soil, good sanitation practices.

•Biological: Insect parasitic nematodes (entomopathogenic nematodes) from the genus *Heterorhabditis*.

•Chemical: Azadirachtin can provide control; used as a drench to the soil, will also control cutworms.

Whiteflies; Family *Aleyrodoidea*

Description: More than 1,550 known species comprises this difficult to control pest. A vector of plant diseases. Feeds by tapping into the phloem of plants with damaging toxic saliva. The pest congregates in large numbers and causes further danger to plants from a mold that grows on the whitefly secretions of a honeydew like substance.

Distribution: Widespread; a problem in greenhouses.

Symptoms of Injury: Scarified leaves on new, tender growth occurring on top or underside, blackened powdery substance from heavy infestations.

Controls

•Cultural: Scout for pests with sticky traps and begin control immediately if detected. Populations increase rapidly, and early control is critical to prevent significant economic damage. Minimize plant exposure, quarantine new stock. Hand removal of leaves heavily infested. Water sprays (syringing) may dislodge adults. A hand-held, battery-operated vacuum cleaner to remove adults off leaves may assist in reducing populations.

•Biological: Parasitoid, *Encarsia Formosa* (beneficial wasp) usually effective against greenhouse whitefly, *Trialeurodes vaporariorum*. Beneficial insects including Lacewing larvae.

•Chemical: Neem oil, Sesame oil, Pyrethrum, Capsaicin, insecticidal soaps labeled safe for food crops.

Thrips; Order *Thysanoptera*

Description: More than 5,000 described species; tiny, slender adult with fringed wings. Some species that feed on mites are beneficial while others that feed on plants with commercial value like cannabis are pests. Thrips may serve as vectors for plant diseases such as Tosoviruses. Thrips pupate in the soil.

Distribution: Widespread; hosts are broad and may vary within species.

Controls

•Cultural: Scout for pests with blue sticky traps. Control difficult and complex. Remove infested stock from grow areas.

•Biological: *Verticillium lecanii* (beneficial fungi) may affect eggs, larvae, and adults when used as directed. *Beauveria bassiana* (beneficial fungi), beneficial nematodes (attack pupae).

•Chemical: Neem oil, insecticidal soaps, Pyrethrum, insecticides labeled for food crops.

Other damaging pests to cannabis plants with regional distribution include Aphids, Budworms, Cutworms, Earwigs, Flea Beetles, Leaf miners, Leaf hoppers, Slugs, and Snails. Control methods include cultural, biological, and chemical; use food-safe pesticides that specifically lists these pests for effective control.

RODENTS

Pocket Gophers; Family *Geomyidae*

Description: Simply called gophers, these burrowing rodents are very destructive pests to marijuana plants in the ground. Attracted to any irrigated plant, gophers will destroy a cannabis plant from the root system consumption and may even pull plant material into their underground tunnel system. Characterized by distinctive shaped mounds resembling a volcano, dispatch gophers through fumigation, or trapping to protect cannabis that is not growing in protective cages or not hardware cloth protected. Extensive tunneling systems may cause irrigation waste or expose roots to dry conditions and necessitate pest control efforts.

Distribution: Widespread.

Controls

•Mechanical: Traps.

•Chemical: Rodenticide baits, approved for use around cannabis like Capiscum Oleoresin, Putrescent Whole Egg Solids, Garlic, etc.

Moles; Family *Talpidae*

Description: Moles are small mammals that live underground, with a diet of earthworms and grubs. They are not a threat to cannabis plants however tunnels near the soil surface to catch earthworms may disrupt irrigation systems or causes unwanted escape routes for irrigation water from basins around plants.

Distribution: Widespread.

Controls

•Mechanical: Traps.

•Chemical: Commercial baits labeled for control of moles; although difficult to poison.

Other wildlife that may cause damage from foraging or indirectly from digging around plants because of an attraction to moisture or fertilizers include Deer, Rabbits, Squirrels, Skunks, Opossums, Raccoons, Rats, and Mice. Each animal requires different methods for protection of cannabis plants, and your local plant nursery or hydroponic supply store can offer advice and products for control.

The eradication of any pest that is damaging crops is an action that requires careful thought and precise implementation. Laws protecting species and environments must be the guide, from acceptable products and methods to proper and lawful application, along with the disposal of waste that is a by-product of the process. Premium cannabis farmers must always remain mindful of their ecological responsibilities and the repercussions of every action they take in pest management for optimal crop production, despite where they grow.

Neem

A member of the mahogany family, Neem is a tree native to the Indian subcontinent, but distribution is widespread. Azadirachta indica produces fruit and seeds that are the source of Neem oil, a commonly used agent for pest and disease control in cannabis cultivation for three primary reasons; it is effective, economical, and ecofriendly.

Several products derived from Neem trees range from toiletries to valued honey from bees that obtain nectar from flowers on the tree. Cosmetics, soap, and fertilizers made from Neem by-products are other examples of the beneficial uses from the trees that are numerous.

For cannabis growers, Neem is a useful agent as a natural alternative to synthetic pesticides, especially for organic growers. While it does not immediately kill pests on contact, it serves as a repellant and acts as an anti-feeding protective against pests that usually die in twenty-four (24) to forty-eight (48) hours from starvation. Additionally, Neem oil suppresses insect and arachnid eggs from hatching. Neem based sprays disable adult pests like whiteflies and will significantly reduce dispersal and successive generations. Regular application at ten (10) day intervals provides the best control of insect and mite pests common to marijuana plants.

Neem is available as an oil, or as an ingredient in other commercial products that are effective pest managers, like Azamax®. When applied per the instructions with rates that are suitable for the plant age and pest infestation, Neem and Neem based products are most useful in controlling enemies of a cannabis crop and should be part of a premium grower inventory of pest control products.

Vigorous cannabis plants are living material that is vulnerable to the laws of biology. Great harm can come to your cultivated plants from pests and disease, causing damage to the quality and yield of the successive crop. Prevention is the name of the game and quick treatment if infestation or infection occur is critical for protecting your marijuana plants and your investment of time and money.

Any organism that causes harm to a cannabis plant is of great concern to a grower. With the high value of each plant in consideration, damage from pests or diseases can lead to substantial crop loss and sometimes, it can be economically devastating. While many organisms that are harmful attack vegetation above the ground, control for those in the soil is just as important to plant health and essential to top yields.

The best approach to pest control programs is to prevent infestation in the first place. While natural and synthetic pesticides are available for most conditions, there are consequences to pesticide use, despite purported safety claims. Additionally, many effective control agents are also expensive and part of the cost of growing operations. Given the risks, costs, and potential for residue in the final product, treatment with pesticides should always follow good prevention procedures.

STEPS TO PROTECT YOUR CROP FROM PESTS

•Keep cannabis plants together in the same vicinity, outside or inside and as far away from other plant types as practical.

•Prevent animal exposure (including pets, wildlife, insects, etc.) where possible and minimize human contact to authorized growers, only.

•Practice good hygiene; regular hand washing is important and use of gloves when handling growing media or plants should be a standard procedure.

•Wear clean clothing or use sterile clothing covers when tending plants and in use of cultivation facilities.

•Regularly sterilize tools and instruments with Isopropyl alcohol or Hydrogen Peroxide.

•Clean containers and reservoirs with a mild bleach solution before each use.

•Store only unopened soil and amendment products, dry and protected.

•Add beneficial organisms; beneficial fungi and bacteria added to the growing media for reduction in the need for fungicides.

Once detected, preventing the spread of a disease or pest problem promptly is very important to minimize damage and loss. During crop maintenance, thorough examination of plants and growing media regularly can help to stop any disease outbreak or pest invasion, making it part of the operational procedures is very prudent.

An examination should consist of a macroscopic analysis for abnormalities and when possible, a magnified look (14X minimum) at plant leaves and stems.

INTEGRATED PEST MANAGEMENT

The goal of Integrated Pest Management (IPM) is to manage insects, plant pathogens, and weeds with the least effect on agricultural and ecological systems. Keeping pesticides at minimal levels; a broad range of techniques is available including the use of natural pest control methods. In 1972, IPM integrated into applicable agencies within the United States government and by 1979, President Jimmy Carter established an IPM Coordinating Committee that developed practices for implementation based on six parts:

1. Acceptable Pest Levels with thresholds for control, not eradication.

2. Preventive Growing Techniques that focus on healthy crops as a first-line defense.

3. Monitoring with regular observation and record keeping.

4. Mechanical Controls to control pests like tilling, traps, barriers, etc.

5. Biological Controls for effective control at lower cost like using beneficial insects.

6. Responsible Use of pesticide only as required and must reach intended targets.

Note: Companion planting, often used in IPM, is not practical in cannabis cultivation.

CROP SANITATION

Disease prevention and management strategies should revolve around keeping plants as healthy as possible and that starts with good hygiene standards for handling plant material. Simple procedures will help protect your valuable cannabis plants, from seed to flower.

•Quarantine

Keep any new plant stock introduced into your growing operation isolated from other plants until you are certain they do not harbor pests. That includes cuttings or clones from dispensaries or retailers. Considering life cycles of mites and white-flies, ten to fourteen (10-14) days are necessary to keep plants isolated for regular examination and treatment if required.

Plants that exhibit abnormalities of any kind during a growing cycle are dangerous to remaining stock. While symptoms may be an indication of a water or a nutrient issue, the danger from pests or disease conditions is too great to ignore. Early diagnosis of disease in individual plants can save an entire crop before spreading, so if possible, identify and treat the pest or condition immediately or else remove the suspected plant(s) from the population of other plants until the problem is no longer present.

•Gloves

Professional growers all use gloves when handling anything that contacts plants; that includes growing equipment, media, and tools.

•Sterilization

A clean, sterile environment is the safest condition for growing cannabis; disinfect facilities and hard goods regularly with Hydrogen Peroxide or cleaning products suited for safe use around plants. Some products are not safe to use near plants and suited for hard surface sterilization, only. Read labels to be certain of the intended use.

Any tool used in cultivation, including shears for pruning or grooming and tools for soil work requires disinfection regularly with bleach, hydrogen peroxide, or alcohol.

•Disposal

Removal of diseased plants is vital for survival of your remaining plant stock. Once removed, diseased, or plant material suspected of disease should not go in the ground, composted, or reused in any way; discard safely. Materials that were in contact with the diseased plant require sterilization or discard, including growing media.

In the interest of plant health and disease prevention, discard growing media that cannot receive sterilization. Crop rotation is important as cannabis should never grow in the same soil or area of previous in-ground plantings. Pathogens in the soil may build to dangerous levels over time and successive planting is dangerous for crops of most types of cannabis.

PESTICIDE GUIDELINES FOR CANNABIS GROWERS*

In Spring of 2015, The State of California, State Water Board, took unprecedented steps to help marijuana farmers use pesticides safely. While not endorsing marijuana cultivation, the guidelines were "for informational purposes, only." While the guidelines apply to California, the objective, and intent is applicable in other locations where marijuana grows under regulation. From the public record, they include:

1. No illegal Mexican Pesticides. The pesticides used must be registered by both The United States Environmental Protection Agency (EPA) and the California Department of Pesticide Regulation.

2. Caveat Emptor; Let the Buyer Beware. There are no pesticides registered specifically for use on marijuana and thus none have been reviewed for safety or human health effects.

3. No Residue. Use only products that contain an active ingredient that is exempt from residue-tolerance levels or are exempt from registration requirements as a minimum risk pesticide under FIFRA rules.

4. Read and follow label instructions carefully. The label defines the law and proper use.

5. Permit Required. If you use pesticides to a field, you need an operator's identification number from your County Agricultural Commissioner. No identification numbers are issued to jurisdictions that prohibit marijuana cultivation.

6. Avoid "Restricted Use" pesticides. Permits will not be issued to marijuana cultivators or their sites.

7. No Restricted Materials. Restricted Use pesticides are limited to use by certified applicators, or to those under the supervision of a certified applicator. Permits will not be issued for marijuana cultivation sites.

8. Protect your workers. Employers must protect their employees from exposure to pesticides by following the label and providing protective equipment and clothing.

9. No rodenticides. Rodenticides that are designated as California Restricted Materials cannot be used; and those that are only designated as federally Restricted Use products can only be used by a certified commercial applicator. There are some rodenticides labeled for below ground applications that are not designated as California Restricted Materials or federally Restricted Use pesticides that can be used if consistent with the label.

10. Use natural rodenticides; Capiscum Oleoresin, Putrescent Whole Egg Solids, and Garlic.

11. Read the State's full legal pest management practices for marijuana growers in California.

(Source: Public Record: CA State Water Board, LEGAL PEST MANAGEMENT PRACTICES FOR MARIJUANA GROWERS IN CALIFORNIA: PESTS OF MARIJUANA IN CALIFORNIA, 2015)

Also in the guidelines from the CA State Water Board:
PESTICIDES "NOT ILLEGAL" FOR USE ON MARIJUANA BY CALIFORNIA**

Active Ingredient with Pest or Disease Treated

(Active ingredients that are exempt from residue tolerance requirements and either exempt from registration requirements or registered for a use broad enough to include use on marijuana.)

Azadirachtin - aphids, whiteflies, fungus gnats, leafminers, cutworms

Bacillus subtilis - root diseases, powdery mildew

Bacillus thuringiensis (subspecies *aizawai* or *kurstaki*) - moth larvae (e.g., cutworms, budworms, borer)

Bacillus thuringiensis (subspecies *israelensis*) - fly larvae (e.g., fungus gnats)

Beauveria bassiana - whiteflies, aphids, thrips

Cinnamon oil - whiteflies

Gliocladium virens - root diseases

Horticultural oils (petroleum oil) - mites, aphids, whiteflies, thrips; powdery mildew

Insecticidal soaps (potassium salts of fatty acids) - aphids, whiteflies, cutworms, budworms

Iron phosphate, sodium ferric EDTA - slugs and snails

Neem oil - mites; powdery mildew

Potassium bicarbonate; sodium bicarbonate - powdery mildew

Predatory nematodes - fungus gnats

Rosemary and Peppermint essential oils - whiteflies

Sulfur - mites, flea beetles

Trichoderma harzianum - root diseases

** *(This list from public records is for reference purposes and updates may occur after publication of this book. Check current listings for accuracy and local listings for applicability.)*

PESTICIDE USE REQUIREMENTS

Under law, pesticides require registration by both the US Environmental Protection Agency and your State. Local Agricultural Commissioners typically issue operator licenses for application, and receive reports about pesticide use. The laws and regulations pertaining to pesticide use on marijuana (like food crops) concern both in ground cultivation and container grown, inside or outside.

•Restricted Use Pesticides

The US EPA designates some pesticides as "Restricted". Any Restricted pesticides are not for marijuana cultivation and should never be in use.

•Rodenticides

Although commercial rodenticides are highly regulated and applied by a certified applicator, they are not safe for use around marijuana cultivation sites. For safe and approved use in most jurisdictions, consider repellent products like Capiscum Oleoresin, Putrescent Whole Egg solids and Garlic.

•Storage

ANY pesticides must comply with State regulations regarding storage. NEVER store any chemical (pesticide, fertilizer, or soil amendment) near a riparian setback or where there is a danger of pollutants entering waterways by accidental leak or discharge. Do not store oxidizing fertilizers or those that are Nitrate based near fuels or solvents. ALWAYS follow label directions and keep all chemicals (pesticide, fertilizer, or soil amendment) away from pets, livestock, and children.

An effective approach to controlling pests and diseases should include a plan to prevent infestation alongside safe pesticides to use for eradication. The more detailed and specific your regimen can become will serve your goals and objectives for premium crops that meet pesticide safety regulations, now or later.

Hygiene & Sanitation
best practices

•Use rubbing alcohol (70-90%) to sterilize all cutting tools and instruments.

•Use a solution of one (1) part bleach to ten (10) parts water to sterilize growing containers.

•Use gloves (pH balanced) for any contact with living plant material.

•Use protective clothing and eyewear when applying any pesticide.

•Use rubber gloves when mixing nutrient solutions.

•Lab coats offer protection to plants from street clothing that may harbor pests.

•Hand washing, before any procedure involving your plants is necessary.

•Filter intake air with filtration or sterilization devices.

•Wear disposable shoe covers when entering indoor gardens.

•Minimize human entry into grow spaces and keep all pets, including dogs, out.

•Keep a spray bottle of hydrogen peroxide handy for quick cleanups.

•Dispose all organic material from the garden appropriately, away from plants in cultivation.

•Do not store chemicals like pesticides and fertilizers near the grow space.

•Discard any empty product containers and recycle if possible.

•Keep grow spaces and gardens immaculate and weed free.

•Do not reuse soilless grow media; instead recycle it on the farm or garden.

•Shield blades with cardboard when discarding disposable scalpels and tools.

From planning and purchasing, to installation and retrofit improvement, all operational considerations should center on conservation of resources. It is a sound way of farming and makes good business sense. Incorporate the theories and practices of successful cannabis growers and find a map of stewardship to the planet.

There are practices that successful and responsible farmers have used for centuries in the protection of agriculture's most important resources; soil, and water. Whether a small indoor grow or a large outdoor garden, soil, and water conservation practice and proper management of these natural resources is critical for premium crops in the short-term and sustainability of healthy environments for farming in the long term. For all cannabis growers, the preservation of natural habitat and processes that support a healthy ecosystem should be a priority.

SOIL MANAGEMENT OBJECTIVES FOR ALL GROWERS

Soil conservation may seem like a complex topic, but the concept is one of common sense and basic protection of the land. The goals in soil conservation aim to prevent erosion and harmful chemical alteration. Responsible use of the soil is achievable with a plan of operational standards for handling your growing media, despite your size, or cultivation methods. The components of a good soil management plan must mitigate any erosive actions and prevent overuse or chemical contamination. Protecting the soil integrity for future crops and future generations is every grower's responsibility and the benefits; both economically and ecologically are huge.

Techniques to maintain soil integrity on or near your grow spaces:

•Crop rotation: Avoid planting successive crops in the same soil; change locations and plant new strains when possible.

•Cover crops: Plant cover crop seed mixes between growing seasons and till before planting.

•Conservation tillage: Till soil when conditions are best and only when necessary; weather plays a role in determining when to till, factoring moisture content and wind. Consult experts for contour tilling and planting recommendations on slopes; seasonal restrictions may apply.

•Prevent runoff: Irrigate to provide only what plants require and eliminate waste. Prevent any runoff into natural habitats or waterways.

•Utilization of living organisms: Earthworms, beneficial fungi, and beneficial bacteria all play an important part in soil improvement and help to achieve conditions optimal for plant health.

•Mineralization as needed: Apply only what is necessary to correct deficiencies for cultivation and avoid any over-fertilization; less is always more when it comes to application of fertilizer products for the soil.

WATER CONSERVATION

The quality of water for irrigation of cannabis plants is maybe one of the most influential factors in cultivation. Filtration of impurities and chemical correction of pH values is necessary and required for best results; however, there are consequences from the process effecting soil conservation and must be part of conservation efforts for both resources.

A disposal plan for waste water from reverse osmosis filtering is an example of why it is important to consider not only water sourcing, but also the discard of any waste water used in the growing operation. In hydroponic systems, reservoirs, and supply tanks change regularly according to the nutritional program. Replacing the depleted nutrient solution and biological material from the media or plants every seven to ten (7-10) days requires disposal. Surface dumping is not acceptable in soil conservation and may cause serious harm to the environment and your growing operation, especially if it contains pathogens or harmful chemical concentrations.

DRAIN TO WASTE

Cannabis farmers who grow in containers often use nutrient solutions applied to the surface of the soil or growing media, in the same way as irrigation and allowed to penetrate through the roots to drain out the bottom. These systems are "drain to waste" and identify another source of waste water to consider when planning a site for growing.

Commercial growers typically employ filtration systems to recycle waste water for reuse. While cost factors may prevent small growers from utilizing such a system, alternate methods exist. Most are regional solutions with regulations that your local agricultural extension office can inform you about.

DISPOSAL OF USED SOIL OR GROWING MEDIA

A thorough soil conservation plan should include a safe and practical way to dispose of spent growing media. Although growing media may appear to be suitable for reuse, aside from nutrient depletion, and physical changes that alter the characteristics of it, soil or soilless growing media used for growing cannabis even one time may contain pathogens that would be harmful to successive growing in the same material.

Usually, the organic soil materials from growing cannabis are safe to integrate into outside, native soils for improvement of those soils, unless there was a disease condition that effected plants. Inorganic materials and those suspected of harboring pests, or disease requires a disposal according to local agricultural recommendations.

Composting is a common practice among cannabis growers who chose to recycle their spent growing media and soils. To complete the process properly, efficiently, and for the good of anything you wish to use it upon, compost must breakdown with heat, water, and air. The natural process not only changes the physical structure of the material, but also the chemical.

Simple compost bins are available and easy to assemble from kits or construct by design for recycling and reusing your soilless medium, or you can purchase models from farming and garden supply companies. Essential to composting is heat from the sun, air from turning, and moisture to promote biological processes that break down the material. Take care not to save for reuse or composting any root or plant material that was diseased or is a suspect for disease. The last thing you want to do is introduce pathogens to your subsequent plantings and the reason so many cannabis farmers do not reuse any soil or substrate.

Conservation is not a buzz word but a way of life for any responsible grower. Good farmers of any crop take good care of their land and water, but exceptional farmers take it to the level practical for both excellent stewardship for premium production and sustained to use, generation after generation. Farmers of premium cannabis should be the latter.

Water Storage

Saving water for a rainy day sounds odd, but it is a prudent thing to do when you have living plants to worry about. Water supplies are often dependent on external forces like utility companies, electric pumps, pressure issues and the like. Despite your water source, even if delivered by truck, you need to plan for storing water; it is some of the best crop insurance that you can buy.

Cannabis hybrids in cultivation are not drought tolerant, and while they can endure a short time without water, lengthy durations will lead to crop diminishment or worse, plant failure and death. It is very wise to plan for a scenario when you have no water on demand from your usual source. For many cannabis growers that means tanks or reservoirs for water storage.

The market is wide and deep with storage solutions for every farming need. Both indoor and outdoor growers can choose from portable or permanent storage in nearly any capacity you might require. Water is heavy, and water is dangerous for drowning or electrocution, so make certain that your system and method are safe, and approved for water storage.

An excellent source for water storage tanks, reservoirs, and plumbing supplies are at hydroponic stores and farming supply companies. They might not have what you need in stock but have access to many distributors in the water storage business. These sources are also great places to obtain planning and installation guidance.

A common dilemma when planning is determining the right amount of storage required to have an adequate reserve if usual supplies are not available. Minimally, a 72-hour supply for your plants at peak use is a good target but it will depend on your budget and space availability. Some growers use multiple reservoirs or tanks for a complete system, while others use tanks for topping off hydroponic systems, or as mixing vats for nutrient solutions. Ask for suggestions from your supplier about what might be best for your situation and plan for expansion as your needs may change.

Chapter 6
Water Resource Management: Making Every Drop Count

There are seemingly countless things that can influence the health of a living species in culture, but few can compare with the effect that water has on plants. Since regular irrigation occurs in cannabis cultivation, water, and the role it plays in managing optimal substrate conditions are important to understand for the serious marijuana farmer.

Fortunately, the technology of irrigation equipment and supplies allows growers to use their water wisely and economically; more important advantageously for their plants. This is great for cannabis farmers who require precision in sourcing and distribution of suitable water in their quest for premium bud. Smart water use for a marijuana farmer is a big part of the success equation.

Even with technological advancements, replicating nature's ways should be part of every grower's regimen. Water quality, quantity, and timing of application are attributes that must be in balance, just like in nature, for soil, or substrate that is ideal for plants to thrive. The focus of this Chapter is about water for irrigation of cannabis in cultivation and the role it plays with soil, vigorous plants, and a resulting premium crop.

6 | a. Supply and Waste
The Flow in Water Planning

Water is the life blood of a cannabis farm, and irrigation is a big operational part of cultivation. Good design, proper installation, and regular maintenance are priorities for the benefit of the plants. There are components for any conceivable need in growing marijuana and for any size project. Digital monitoring and controls are very affordable, and new automation equipment appears in the marketplace frequently. The most efficient use and disposal of water must be a primary concern if growing premium cannabis is a serious endeavor and the key to smart planning.

The resources that effect the success of your crop also has a significant effect on the economy of your operation and the ecology of your location. Important systems like lighting, heating, air conditioning, and ventilation consume energy that smart cultivators try to make as efficient as possible. Another vital resource to cannabis growers who require efficiency is water; for either indoor or outdoor cultivation. Just as important as other essential utilities, water is a managed commodity.

Water sourcing is a critical element in planning and operation. The quantity of water required to grow cannabis varies according to the climate (controlled or natural) and method of cultivation. The type of soil or growing media also determines water use. The quality of the water is just as important to cannabis cultivation for those seeking a premium crop. Most water from domestic sources or wells will need some modification (filtering or treatment) to make it suitable and beneficial for use in cultivation.

Non-permitted water diversion from any creek or stream for cultivating cannabis is unlawful and for very good reasons. It limits the water available to the public and to wildlife. With recent devastating droughts like those in the western United States, it is a criminal behavior subject to Federal and State penalties, including imprisonment and heavy fines. Essential point; do not do it.

DOMESTIC WATER AS A SOURCE

Potable, or water safe for drinking and public use is generally safe for plants. Water originating from a well or from a treatment facility that obtained it from groundwater, or other natural water source, tested and deemed safe for human consumption will supply much of the water for cannabis cultivation for readers of this book. For that reason, consider the types of treatment and what effect they have on your living plants.

•Softened Water

Domestic water that has been "softened" can harm cannabis, especially when grown in containers or in hydroponic solutions. Water softeners typically replace calcium with sodium. Calcium is an essential nutrient to plants, while sodium (salt) is not and can be harmful to marijuana plants in concentrations. Correction of the issue involves sourcing the water before softening, or by adding calcium back to the softened water. Add gypsum (calcium sulfate) available at most nurseries and gardening centers to soil at the suggested rate if there is no calcium added to the softened water before application to plants. Liquid forms of supplement for deficiencies include calcium formulations, usually also containing magnesium and iron are a supplement to filtered water just before application to cannabis plants. All hydroponic stores carry forms of this product, often called "Cal-Mag."

The addition of agricultural lime will assist in raising a soil pH that is too acidic or subjected to long periods of watering from softened water. Use at suggested rates.

•Treatment by Chlorination

The addition of chlorine (Cl2) to water will kill microbes and bacteria and may prevent the spread of waterborne diseases. The addition of chlorine or fluoride into water to help purify it is very common, but it will affect your cannabis plants. Removal of these materials prior to irrigation is possible with filtration designated to specifically remove chlorine, or by adding liquid stabilizers (used for aquariums) to water treating tanks or reservoirs. Hydroponic stores and distributors offer products that will filter or prepare water for plant use; consult retailers or your water agency for specific advice applicable to your specific location.

•Treatment by Fluoridation

Fluoridation is the addition of one of three compounds of fluoride to drinking water for preventing tooth decay. The forms of fluoridation compounds used to treat public water supplies is odorless, tasteless, and colorless. Fluoride compounds occur naturally in water at varying levels. A simple faucet filtration system does not alter the fluoride content in water, however reverse osmosis filter systems can remove 65-95% of fluoride, while distillation of water removes 100%. Although a natural element, the effect of fluoride in specific concentrations on cannabis plants is unknown.

•Do Not Irrigate with Gray Water

Recycled water from various sources, including households, called gray water, is not an option for marijuana plants. Despite what you may read in water conservation articles, gray water, while often used in ornamental horticulture or on grass fields, is not safe for cannabis because of a high salt or detergent content, along with unknown contaminants. If you want premium crops, use of gray water will ruin your effort.

PLAN FOR WATER WASTE

Excess water from irrigation, waste water from reverse osmosis filtering and spent reservoir water are just a few of the ways in which you may need to dispose of water when growing cannabis. Indoor cultures or growing setups in greenhouses with solid floors require a legal (by local code) way to drain waste water from the structure, while outdoor or greenhouse facilities with soil floors may rely on containment within the growing area, into the adjacent soil. Despite your irrigation method or location, planning aspects of this important part of water management will be very advantageous for your growing efforts and the environment.

To provide a glimpse on waste water from irrigation, consider two ends of the cultivation spectrum. Growers who utilize in-ground cultivation with drip irrigation have the least amount of waste water per plant grown. Growers that use reverse osmosis water filtering with hydroponic reservoirs of nutrient solution may have a large amount of waste water per plant, as "waste water" is a byproduct of this filtering method and reservoirs of nutrient solution receive change-outs regularly. There are specific reasons for each waste measurement generalization, but the examples illustrate a range. Plan for your specific methodology and get outside advice if you need it.

Growers that use containers and water their plants from the top of the growing medium should allow a 5% to 10% runoff to achieve complete saturation of the root system. A saucer for collection of the water is a technique, and then the water can evaporate, or the water may simply escape into the earth below the container. Use care with saucers; a common reason for failure of marijuana plants is over-watering. Saucers collect and hold water, preventing complete drainage from a root system. Unless the saucers empty by capillary action or evaporation within a few hours of filling, do not use them. Drain-To-Waste fertilizers and conditioners should be eco-friendly if allowed to drain into the grow area soil and as "organic" or natural as possible.

Good water management in cannabis cultivation considers both sides of the important equation; water in and water out. In new construction or remodeling of an existing farming operation, accommodating the sourcing and disposal of water and water-based solutions helps tremendously in resource management and successful cultivation projects.

Managing Water

The importance of water management in cannabis farming requires a basic understanding of the topic and moving forward, an in-depth education about water and wastewater is imperative for any expected success as a grower.

The American Water Works Association, is the first place to start learning about the business of water and how it applies to your operation. Despite your water source; from a ditch, a well, or from a municipal supplier, the rules of the road for water and agricultural use are changing. A highly regulated industry overseeing a finite resource, the world of water is writing new laws with more rules to come. Anticipating those changes and pairing with the water industry at the local level are a worthy strategy for any commercial grower and the AWWA provides the data to guide intelligent direction.

In addition, the website for the American Water Works Association, www.awwa.org, offers information on water and wastewater utility management. The concepts are fully adaptable to use in planning water use, conservation, and sustainable practices for anyone who uses water to grow marijuana for a living. Trends, economic statistics, and more give the agricultural water user in cannabis cultivation useful numbers to manage their operations, plan, and get involved at local and regional decision-making forums or organizations.

Individual States will also offer a wealth of information when it comes to agricultural water use. In California; for example, the State has a very detailed Agricultural Water Management Plan. Cannabis growers who utilize land zoned for agricultural use should consult the specifics for compliance and useful direction from any State or local agency that concerns itself with water.

6 | b. Water Quality
Content and Measurements That Matter

Water quality is a subjective assessment but when it comes to growing premium grade marijuana, it is a big part of the equation for success. The water used to irrigate your cannabis crop directly influences plant health by systemic absorption and indirectly by how it reacts to nutrients that plants need for vigorous growth and bloom. Striving for the best water you can provide based on plant requirements for your specific cultivation method should be the goal.

In the quest for healthy plants that produce premium flowers, cannabis farmers go to great trouble and expense to make certain cultivation factors are as optimal as possible. Measuring and correcting water quality is a critical part of growing premium crops for human consumption, and an effort every marijuana farmer should practice on a regular schedule.

Measurements of the condition or characteristics of water focuses on final use, need or purpose, and references against a set of standards for that use. Those measurements provide data for assessing compliance of rules for water use, safety for a use of water and health of an environment. Water quality is a complex topic with many types of measurements. In agriculture, what is in the water is a very important aspect of any operation.

Alongside the basic need for a reliable source of water throughout the growing stages of cannabis, the quality of the water is also critical to crop success. Measured in several ways, chiefly by biological and chemical analysis, farmers look for anything in the irrigation water that may have an adverse effect on plant health. Contaminants including microorganisms and inorganic chemicals are factors of concern in measuring water quality to marijuana growers and identified in an assessment.

WATER SAMPLING

The metrics used to evaluate water often vary by region, and their applicability is dependent on the final use of the water. Water for domestic use, regulated by various agencies, including the EPA (United States Environmental Protection Agency), must meet strict criteria to be safe for human consumption while water used solely for irrigation may meet another set of standards.

The most accurate measurements are on-site, with subsequent, or comparative analysis completed in a laboratory. Fortunately, the cannabis grower can use hand-held instruments to instantly analyze important factors in their irrigation water and data important to optimum growth; temperature, pH, and electrical conductivity (a gauge of the mineral composition). Recently, real-time monitoring instruments suitable for growing cannabis has gained in popularity resulting from accuracy and affordability, with wireless connectivity to mobile devices.

Samples for laboratory testing should follow protocol set by the lab. This includes where to take the sample, how etc. For analyzing water destined for use on your cannabis plants by hand-held devices like pH and EC meters, sampling should occur at the water output point, not the source, as changes can occur as the water makes it to your irrigation valve and dispersal fixture.

For best results when using meters or testing strips, collect a small amount of water from a sprinkler or hose valve into a bucket, as close to the irrigation dispersal point as possible or if growing hydroponically, directly from a reservoir or mixing tank. When using hand-held meters, a small cup or container is handy for water sampling or the meter probes insert directly into storage containers, reservoirs, or tanks of water for precise measurement. Manufacturers of measuring devices often suggest methods specific to their product and users should read and follow those instructions.

BEST PRACTICE: *Store pH pens with wet tips; if they dry, they die.*

QUALITY INDICATORS

Determining water quality relies on criteria obtained by examining what are termed indicators, or attributes of the water. The water quality indicators for drinking water are measurements of specific contents considered important for safe consumption by humans. Those requirements vary and may be different from indicators useful in water assessment for crop irrigation in the same locale. Additionally, treatment of water for domestic use is often by filtering, fluoridation, or chlorination to meet requirements or standards, resulting in different composition than non-potable water deemed safe for irrigation. Standards vary by region; consult your local water agency, supplier, or agricultural extension office for more information.

Water Quality Indicators important to farmers:

• Physical Indicators: Temperature, EC (Electrical conductivity), color, taste, odor, TDS (total dissolved solids), turbidity (transparency)

• Chemical indicators: pH, Total hardness (TH), heavy metal content, pesticide residue, nitrate, surfactants, and orthophosphates

• Biological Indicators: Aquatic insects including Caddisfly, Stonefly and Mayfly and Bacteria, including *Escherichia coli* (*E. coli*) and Coliform

ADJUSTABLE FACTORS

Growers use many systems for measurement and evaluation with various methods to clean or adjust the water used for irrigation of cannabis plants. Reverse osmosis filters, for example, remove sediment, and contaminants, while also modifying the pH level of water. The requirements for optimal cannabis growth include:

• pH

Range: 5.5 to 7.0, Ideal: 5.8 to 6.2

Adjust with chemical solutions prepared and sold for use. Do not try to mix your own with dangerous chemicals! Buffered, the products prepared for safe use (labeled pH UP or pH DOWN) are very inexpensive.

• EC

Electronic test meters measure EC, then convert to PPM (parts per million)

Range: 500 to 2000 ppm, Ideal: 800 to 1200 ppm

• Temperature

Range: 60°-70°F (18°-24°C) Irrigation water temperature should be near soil temperature to avoid shock to roots. Do not use cold water to cool soil or hot water to warm soil; it will cause damage to your plants. (Note: The desirable temperature range for water used in irrigation is like the ideal soil temperature range for plants.)

The addition of any chemical or fertilizer to irrigation water will change the composition and may alter the desirable range of measurement of adjustable factors.

Always conduct a final test before use or application of altered water to meet safety ranges of pH, EC, and temperature.

WATER TYPES

• Hard Water

If a soil is very alkaline and the water for plants is "hard," the desirable pH range for cannabis is adverse. The addition of iron chelates to planting areas or soil may offer some relief, especially if new growth appears yellowish. Adjustment of an alkaline soil is also possible with products like soil sulfur.

• Distilled Water

One would think that pure water, with all contaminants and minerals removed would be perfect for growing cannabis. In theory, yes, but you would need to add macro nutrients, micronutrients, and beneficial organisms along with various compounds to make it "ideal" for plant consumption.

The best use of distilled water is in foliage sprays. Lacking any salts, there is no residue on the plants or equipment, nor is there any materials in the water that can react with biological components or chemicals used in spray solutions.

When cultivating premium grade cannabis, water supply is a multi-part component of the total farming plan. While the necessary volume is important, the physical, chemical, and biological aspects of your irrigation water require regular measurement, monitoring and sometimes, alteration, to fit the needs of vigorous plants. Keeping the key factors at optimal levels for cannabis helps a great deal in maintaining healthy mediums and achieving bountiful harvests of desirable crops.

6 | c. Delivery Method
Select A Way to Maximize Every Drop

The cultivation method used in growing cannabis determines the ways you can irrigate the plants, but choosing one that delivers the water accurately with little waste is the objective for a good system. Even in hydroponic culture, water delivery must be precise while reservoir storage must be ample. There are several resources available for planning and installation, making a state-of-the-art system both affordable and easy to install.

Methods of irrigation for a marijuana farmer have one objective; deliver a suitable amount of water to the root zone.

Aside from hydroponics where cannabis plants grow in a nutrient solution, there are numerous methods of applying water to the roots of plants; however, some are inappropriate for growing marijuana for consumption. Hemp, a relative to cannabis strains discussed here, employs methods like furrow watering, flooding and even sprinklers to irrigate, but these techniques are not useful to the premium cannabis cultivator.

In cultivation of sativa and indica cannabis strains for medical or recreational use, it is prudent to keep water off the foliage or to a minimum during vegetative stages and to eliminate it completely during the bloom phase. This helps prevent diseases like powdery mildew or *Botrytis* that thrives on damp leaves and stems. Once established, these pests are difficult to eradicate and can wreak havoc on your plants.

Marijuana plants that are in bloom phase are especially vulnerable to pathogens when water collects on and between hairs and trichomes on the buds. So, humidity should remain below 50% if practical in the growing environment throughout the vegetative stage with no water sprayed on or applied to the foliage during the bloom phase of growth. An obvious exception is the application of preventive or treatment sprays to control pests, required throughout cultivation.

MANUAL APPLICATION TOOLS

• Watering Cans

Useful for small cultivation projects, drawing from rain barrels, or for applying water-based nutrients to plants in containers or in the ground, particularly when plants are young.

• Hose-end Nozzles & Wands

Essential for hand watering, nozzles at the end of a garden hose give you the ability to control the flow of water and to shut the water off to prevent waste. Many nozzles also provide a pattern adjustment that provides a way to replicate soft rain. This gets the water evenly dispersed without flooding or rutting from just the use of an open-ended hose.

Extension wands make irrigating containers and in-ground watering much easier by placing water more accurately and thereby reaching the root zone. As plants mature and their drip line expand, wands save your back, conserve water, and increase efficiency. Most have independent shut-off valves that add to the convenience of saving water.

Misters and foggers are not useful to the marijuana grower except to maybe add humidity to seedlings or young plants in a dry environment; a dangerous practice for mature plants. Overhead emitters installed to spray pesticides to foliage are an exception.

HOSES

Garden hoses are a basic irrigation staple, despite your watering system. Selecting the right hose for your application will save you both time and money, besides avoiding frustration, and ineffective watering. The best advice is to avoid, inexpensive vinyl; most kink and have poorly made fittings. Instead, choose rubber (resistant to UV damage and cracking) or a rubber and vinyl combination.

Hoses lengths vary, in 25 foot increments and although it is tempting to purchase a long one ("just in case, for all needs"), you will waste money and require additional storage space for the unused length. It is better to connect shorter length hoses if you need an abnormal length for a special project. Hoses with higher burst strength will stand up to high pressure water supplies, and generally suggested for cannabis garden use.

DRIP IRRIGATION

- Drip systems

Serious cannabis growers rely on precise control of irrigation and drip irrigation systems offer the exactness they need. This method is a popular choice for efficient delivery of water with very little waste. After a moderate initial cost to set up a system, drip or "trickle" systems have several benefits. Water loss to evaporation decreases; water goes to an exact root depth while weeds competing for irrigation water are significantly lower in number.

Estimates range from thirty to seventy (30-70) percent reduction in water usage over other irrigation methods resulting from water provided at a slow and steady rate. Drip irrigation systems are highly customizable with fittings for connecting to solid water pipe, like PVC or copper with numerous adapters and fittings for water distribution. Supply lines, typically 1/2 inch is flexible hose-like and available in 25 feet to 100 feet lengths and in some stores, by the foot. The feeder lines, typically 1/4 inch to 1/8 inch connect to the supply line with barbed fittings and come in similar lengths as supply line tubing. Emitters are available to drip or sprinkle at controlled rates, to complete the distribution of water to the plants.

Drip systems are available by the individual fitting or in kits. Tools and accessories to properly install the systems are necessary, like punches, fasteners, stakes, etc. While most are universal, some fittings are specific to a manufacturer's system, so always check for compatibility. A plastic tool box comes in very handy to sort and store extra fittings and tools for repair or additional installation later.

Useful Internet links for planning and supplies:

https://www.orbitonline.com/
http://rainbird.com/
http://www.raindrip.com/
https://www.dripdepot.com/

- Porous hose watering systems

Termed "soaker hoses," this method uses osmosis to release water through thousands of tiny holes. The hoses must be inside the drip line of a plant to provide ample water without waste from runoff or evaporation. Soaker hose tubing used in drip irrigation systems is easy to use this way, but soaker hoses of large diameters in long lengths are more challenging to use efficiently for individual plants.

AUTOMATION SAVES TIME, REDUCES WASTE

• Irrigation Timers

The convenience, affordable cost, and the protection that plants will not miss a watering are very good reasons to use timers. A battery-operated hose faucet timer is a starting point at $20.00; however, an electrical irrigation timer that start at $40.00 is worth every penny in cost. Most have a battery backup for power failure, and some, termed "smart timers" connect to sensors that automate irrigation by turning on and off various valves. When selecting timers, consider your number of irrigation control valves, or "zones" as the first criteria.

Remote control is an option for higher priced models, starting at $250.00 for an eight-zone model with control from a mobile device from anywhere using Wi-Fi technology. Combined with sensors and automated valves that communicate with the controller, newest models simply replace existing controllers and do not require any router ports for wireless connectivity of the entire system. Components in an automated system must be from the same manufacturer and design to work together properly.

Irrigation timers offer programming options from different start times, start days, frequency, duration, and cycles determined by weather conditions. Check the programmers guide on the product packaging before purchase to make sure your timer offers the features you want and need.

If you are an indoor grower looking for state-of-the-art technology, system controllers that accurately control irrigation, temperature, humidity, and CO_2 dispensing with wireless connectivity are now available at very reasonable costs. For new projects or upgrading existing systems, the time could not be better for this type of equipment.

As water for irrigation becomes a more precious commodity, cannabis growers benefit from new and improved technologies to maximize use. Starting with a system that conserves while delivering proper amounts of water or upgrading to one that does is a step that will serve any marijuana farmer looking to be a smart cultivator on a path to growing premium crops.

Irrigation Info
learn to water wisely

• Two prime causes of crop failure are over watering and under watering.

• The leaf droop at the end of day is a natural cooling mechanism, but check for dryness anyway.

• Smaller containers require more frequent watering than larger pots.

• Drip irrigation emitters must saturate the entire substrate in a pot, not just a corner.

• Never use water right out of a hose; it may scald roots if too hot.

• Anticipate increased water needs during heat waves.

• Provide sufficient water to the soil or substrate, not to the foliage.

• Allow 5-10% of applied water or solution to run through a hard-wall container.

• Provide water at the minimum rate of 15-20% of fabric pot volume per watering.

• Use timers or automatic irrigation equipment, but supplement if needed.

• Use nozzles that aerate the water for increased oxygen availability at root zone.

• Do not use saucers under pots that do not drain or evaporate in a day.

• Never collect water in buckets or reservoirs without covers, secure from children, and pets.

• Filters and back flow valves are essential when using drip irrigation systems.

• Repair leaks to stop waste and to prevent wet conditions that harbor pests.

• Measure and chart water use for planning future needs.

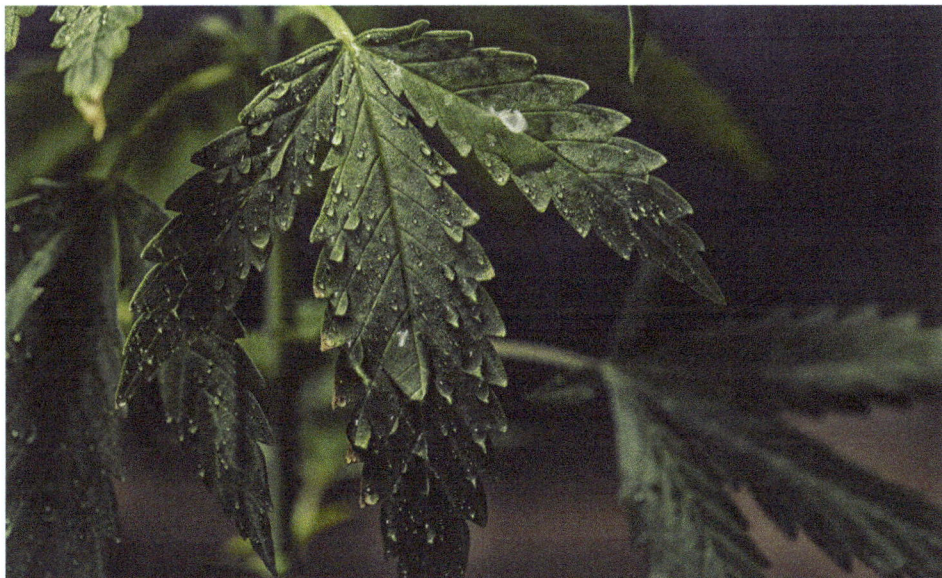

Cannabis plants are finicky when it comes to their water. Some hybrid strains are downright fussy. Too much and they droop from drowning roots, while too little makes them wilt from dehydration. Where watering occurs is just as important as the quantity. Add it all up and it requires some prudence in water application, especially if you want a top-shelf product at harvest time.

Poor watering habits are one of the leading causes of cultivated plant failure. Both over-watering and under-watering marijuana plants contribute to many issues that can lead to harm, permanent damage, and ultimately to crop failure. Plants stressed from too little or too much water, even for one day are vulnerable to pests and diseases, while natural processes that produce quality vegetation and prolific flowering diminish. The right amount of water applied at the right time is critical for plant health and vigor and essential for a premium quality crop, despite the plant type or strain.

Providing the proper irrigation to cannabis plants is not difficult, however providing advice beyond general suggestions is problematic. There are simply too many variables to give specific advice that is reliable. Climate, light exposure, growing medium, and stage of development will all have influences on water demand. Specific conditions of your farm or garden should always be a primary point of reference since every setup is different. Experienced cultivators call this "know your grow," a vital part of success that goes beyond technical recommendations.

It is important to point out that correct watering practices will not make up for insufficient nutrients or poor soil quality. The entire context of information here is with the assumption that soil conditions are conducive to healthy cannabis, drainage is good, and nutrient levels are correct.

Some growers prefer to hand water for what they consider total control, however most choose to use some type of automated irrigation system, and then augment irrigation by hand if needed. This provides convenience and insures that plants will receive the right amount of moisture with regular watering. While irrigation systems and their use have become a complex specialty in horticulture, the basic premise of good systems is the same; efficiency.

WATER SYSTEM PLANNING

Just as important as any aspect of cultivation is water and how you get it to your plants. Before you plant a single seed, clone, or plant, start with some homework, and do some pre-purchase shopping. A visit to stores that stock a wide selection of irrigation systems and supplies will help broaden your horizons and probably eye opening when you see what and how much is available. Most manufacturers offer "how-to" advice for planning in brochures available in the store or from tutorials on their websites. The material selection today is enormous and well-suited for any size growing project.

To do it right, sketch out a plan with plant locations and measurements of distances accurately depicted. Mark water sources like valves or faucets and make certain to plan for drainage, especially with container grown stock or hydroponic reservoirs. Given that water is a precious resource with limits to availability, efficiency in irrigation goes beyond quantity and timing of application. Minimizing waste, or runoff is a vital part of water management and must be part of any irrigation plan. It is smart business, practical farming and good for the environment.

If your system is complicated, use different colors to represent different elements in your plan and make certain to allow a way of expanding your system. As your projects become successful, you will be glad you planned every detail. Finally, save a copy of the final installation (since there may be changes) for future reference.

While irrigation system materials are not expensive relative to their functionality and long-life, planning beforehand helps ensure compatibility of individual components and a material list of parts needed makes shopping much easier. Drip systems are very popular with cannabis farmers with scores of pieces and parts for complete adaptability to nearly any setup. Preparing a shopping list also helps to avoid multiple trips to the store for forgotten items, while purchasing common items like elbows or connectors in "contractor" or multiple unit packs are economical and save getting the exact number when you buy.

Small sprinkler or irrigation systems can be fun to install with a good plan and not that difficult after any trenching or underground work is complete. Larger projects may require the assistance of an irrigation specialist or contractor, especially if installation requires Permits. Primers and glues make PVC connections strong and reliable while drip system parts typically snap together and rely on compression for a seal. With monitors, timers, regulators, and even automatic fertilizer injectors, irrigation systems can be as simple or complex as you choose.

IRRIGATION OBJECTIVES

•Plants are Individual Biological Units

A common misbelief is that cannabis plants take an inordinate amount of water. Each plant requires a different amount of water based on strain, stage of growth, climate conditions and so on. Therefore, there is no set amount of water that plants require; it is purely an individual plant's need. This principle applies to any method of growing, indoor, or out.

So how do you know how much water to provide? When growing hydroponically, a reservoir stays at a certain level for roots and pumping systems, relying on re-circulation of nutrient-enhanced water from it to provide plants what they need. For container grown stock that uses a drain-to-waste method, watering occurs at regular intervals to the soil surface, at sufficient amounts to wet the growing soil or media, with slight runoff of about 10%.

In container grown stock, irrigation frequency may be daily, twice daily, twice weekly, or whenever the soil is semidry or the plant expresses a need, typically by wilting of new growth at stem tips. Some wilting occurs naturally if mature leaves become warm and the plant slows transpiration, a cooling mechanism that occurs at night. Over-wilting from lack of sufficient water may cause serious and permanent damage to plant tissue; avoid at all costs. (Over watering may also lead to wilting because of a lack of air in the root zone.) As a rule, frequent monitoring of plants is a good

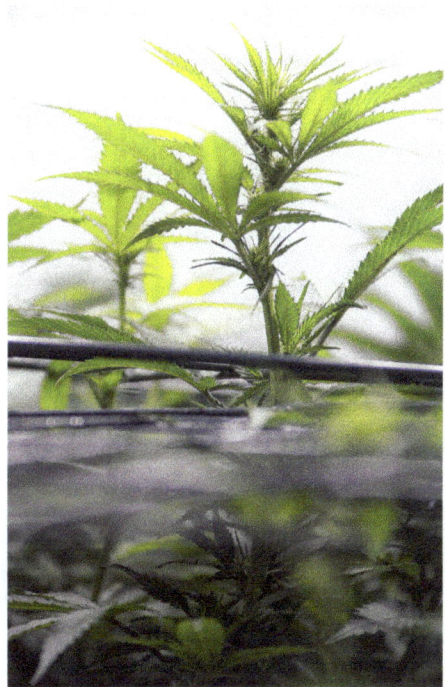

practice, throughout the growing and flowering stages, to avoid stress to your plants caused by water issues.

Marijuana plants growing in the ground have a completely different regimen for irrigation than hydroponic or container grown stock. As a rule, wet the root zone, and let dry slightly before the next water application. Soil types, and planting hole preparation plays a huge role in water required for proper irrigation of in-ground plants, not to mention the sun and climate conditions. Again, knowing your specific soil characteristics is essential for good watering practice.

IRRIGATION TIMING

•Determining When to Water

A growing medium or soil should never dry out completely between water applications (growers call this bone dry). Experienced growers can determine when a soil or medium needs water simply by touching it. If it feels moist or clumps in your hand, irrigation is likely not required. If the substrate appears lighter in color than when moist or particles do not clump when squeezed or do not stay clumped when you open your hand, the soil likely requires irrigation.

In coco based growing media, it is more challenging to determine water need in a physical touch or clump test. Fortunately, long strand coco fibers have the capacity hold a large amount of water and make it a desirable media for growing cannabis; so, water irrigation frequency is less over traditional soil mixes.

Tensiometers, also called soil sensors, are electronic devices that measure the moisture content in media or soil and most useful in getting to the root level where the water content is most critical. Automated sensors detect data and instantly change the application or flow rate of water by controlling valves. Soil sensors available today are good tools to know when and how much to water, as well as in reducing waste of water and nutrients.

Especially useful in container growing, some tensiometers give more accurate readings than others, so compare products before purchase. Irrigation suppliers and specialty stores can provide pre-purchase advice that is helpful for selecting the right instrument for your specific application. The instruments utilize a long metal probe that inserts into the growing substrate or into the ground within the approximate root zone. Tensiometers that measure in centibars is the most accurate. If the meter indicates 50 to 70 centibar range, it is moderately dry, requiring irrigation. A reading of five (5) to fifteen (15) centibars indicates adequate moisture for the plant.

A low-cost moisture meter gauge is not precise but rather shows water content "ranges" of dry, moist, or wet. This can be of some use if you are on a strict budget and keenly aware of when your plants need water and where the needle or pointer on the device is when this occurs. Use that mark as the basis for a dry substrate requiring irrigation for future reference.

•Irrigation Time of Day

In agriculture of any type, working with nature by replicating ideal natural environments and conditions produces the best result. You will read and hear many theories of what time of day to water for best results. The best advice is to water in the early morning to reduce evaporation and incidence of plant disease.

Avoid watering at night or in darkness, especially outdoors; it encourages the growth of pathogens like mold and mildew except in very dry climates or conditions. Watering at night also invites damaging wild nocturnal animals including skunks, opossums and raccoons, a challenge for outdoor growers.

ADDITIONAL TOOLS FOR IRRIGATION MANAGEMENT

•Meters

A flow meter designed for water accurately measures how much irrigation water passes through it. This is very helpful when you fine tune your irrigation schedules, or for simple monitoring to reduce unnecessary watering (waste). Records of flow measurement are also valuable for planning anticipated needs for future growing projects.

•Mobile Apps

Irrigation management apps for mobile devices allow users to monitor and control irrigation based on changing environmental or climatic conditions. A search of app stores shows what is currently available.

Companies that offer a range of products for the ultimate in monitoring and control also manufactures equipment for auto dosing of nutrients and pH correction based on sensor readings, timers, or schedules. Two sources include Bluelab®, the manufacturer of a popular pH pen (www.bluelab.com) and Growlink® (www.hydropods.com).

Irrigation system efficiency is a ratio of the output to the input, but in practical terms, it is about managing water, time, and effort. Too much water, and your plants will suffer from a lack of air at the root zone. Too little, and they will wilt and the risk of severe damage to plant tissues. The skillfulness in delivering the right amount of water coupled with the right time will not only save water, but serve as a wise economical move in the pursuit of top-shelf flower buds.

Despite your location, there is a good chance that your water use is under regulation or monitoring in some way. How you store your water is also likely under control by codes and restrictions. The key to managing your water source and remaining compliant is to know or estimate your irrigation requirements at the peak of the season and then plan accordingly. The very last thing you want to lose is water, and some jurisdictions will terminate or limit service for non-compliance, so know the rules and follow them exactly.

Moonlight Gardening

Theories about growing plants according to the phases of the moon have been around since humans began farming. Early scientists dismissed the concepts as superstitions, but there is some science behind when to plant crops as influenced by the light and gravitational pull of the moon.

Cannabis growers are always looking for methods of enhancing the cultivation of their plants. Gardening by the moon may be yet another way of working with natural cycles to produce premium crops. Current thinking in science is that the gravitational pull of the moon influences the moisture in the soil much as it does the tides, to such a degree that it affects the seed and root response; especially following a New Moon, while moonlight affects the leaf and flower development.

Farmers who use the garden by the moon system contend that by the second quarter of the moon phase, the pull begins to lessen but the light is strong, promoting leaf growth. Calendars on the subject indicate the water absorption by seeds is highest at the full moon. As light diminishes after a full moon, the energy decreases although the pull remains high making it a favorable time for transplanting. Farmers who subscribe to the concepts use the third quarter for pruning benefits.

By the fourth quarter, the light and pull decrease and growers think of it as a resting period for their plants.

More information on farming by the moon phase is available; the Farmers' Almanac® offers a 'Gardening by the Moon Calendar', with specific activities based on both the phase and position of the moon. The information is applicable to all zones and worth subscribing to with other agricultural information included. The first few months are often free; subscription is very nominal.

https://farmersalmanac.com/calendar/moon-phases/

Plant cannabis seedlings and clones between the light of the new moon and the full moon.

Chapter 7

Light: A Vital Factor for Plant Growth and Bud Development

Like most plants, cannabis requires the correct light to develop and mature. A marijuana farmer can directly influence plant growth by using light for maximum benefit. For indoor growers or even outdoor growers who start their plants indoors, an understanding of light for marijuana plants is advantageous.

Certain bands of color in the light spectrum send important signals to plants. In marijuana cultivation where rapid growth is desirable, these bands and their intensity influence the vegetative and bloom stages of development differently. With proper lighting that provides the ideal spectrum for the right stage of development, plants benefit with not only faster growth, but with better crop yields as well.

Grow lighting for indoor use offers many choices and price ranges. The goal is to look for the most efficient use of lighting relative to cost and operation. The "best" is what works for your cultivation method and size of your garden, so choose the most advantageous type for plants at various stages of growth, but also keeps the electricity expense sensible during use, relative to the profit, or return of finished product. Balancing plant benefits, the initial cost, heat production from the bulbs, and total energy use during operation are the principle determining factors for fixture selection and this chapter compares options on those criteria.

7 | a. The Effect of Light on Crop Production
Optimize and Control for The Best Plant Response

All cannabis growers eventually require an understanding of how light influences plant growth, even if they grow outside. Seeds and clones usually start indoors, and many growers complete growing cycles under artificial light. Knowing the basics about spectrum, intensity, and duration help to gain the maximum performance at each stage of plant growth.

Light plays a significant role in the cultivation of premium grade cannabis and working with natural or artificial light are part of the cultivation process. The source, the spectrum, the timing, and duration all effect the rate and quality of vegetative growth and subsequent blooms. Working with the sun's natural light or with an artificial source, the art of cultivation becomes refined and very beneficial when mastering light.

In the simplest terms, light is bands of color (a part of the electromagnetic spectrum). Different light produces different spectrums that send various signals to plants, affecting each stage of growth. Not all colors in the spectrum are important to plants. Those most influential are in the red (beneficial to bloom phases) and blue (beneficial to vegetative phases) range of colors, and lies in the 400 and 700 nanometer range of light measurement; called the zone of Photosynthetically Active Radiation (PAR). PAR watts are the measurement of photons required for growth. Without getting too technical, these are parameters you should become familiar with to provide a cultivation plan and if growing inside, a guide for light and fixture selection as well as timer settings once installed.

Light sources influence light color. The term Correlated Color Temperature (CCT) is in degrees kelvin (K). This measurement of when the bulb is completely hot and is in full glow is a rating of the bulb. The color temperature of the bulb is useful; however, it is the lamp's Color Rendering Index (CRI) that provides the bands of color represented by the light source. A higher CRI rating indicates a better quality of light distribution for cultivation.

PHOTO PERIOD DEFINES TIMING

The relationship between light periods and dark periods during plant growth in a 24-hour period is the photo period. Cannabis plants are in a vegetative stage with 18-24 hours of light, while a period of 12 hours of darkness per 24 hours (inductive photo period) will begin a bloom stage. There are exceptions, including auto-flowering strains that begin blooming by chronological age of the plant. Also, genetics play a role in light response, so each strain will respond to light in different ways. For example, indica or indica dominant strains typically will respond and bloom sooner than sativa strains.

LIGHT INTENSITY

Light intensity is the key factor in photosynthesis of plants, the "fuel" for cellular growth in plants. Just as spectrum and timing are key factors in growing premium quality marijuana, so is the intensity of the light. If you want to grow dense flower buds, adequate growing light is essential; dark, dim, or shady conditions will produce light and airy buds, often called "popcorn". For our purposes, light intensity is the energy per unit of area, and as light moves away from the source, the intensity decreases. So, generally, the closer a cannabis plant is to the light source, the more PAR watts it receives, resulting in more rapid and vigorous growth. Maintain a proper distance so heat from the lamp does not cause harm to the foliage.

It can be advantageous to use more, lower-watt lamps than one large-watt lamp, primarily because they will cover more space and can be closer to plants. This will provide more intensity and deeper penetration into the plant centers. The initial setup

9000-10,000 °K	Shade
6500-8000 °K	Overcast or Foggy
5000-6500 °K	Full Sun
5000-5500 °K	Photo Flash
4000-5000 °K	Fluorescent Light
3000-4000 °K	Clear Sky
2500-3500 °K	Incandescent Bulb
1000-2000 °K	Candlelight

KELVIN TEMPERATURE SCALE

In physical sciences, the kelvin is the primary unit of measurement. At 0°K, a hypothetical temperature, all molecular movement stops. 0°K is equivalent to -273.15°C, or -459.67°F

cost is more with this approach, and it is slightly more expensive to operate, but the results are typically higher yields of better quality. Follow manufacturer guidelines or ask when purchasing about coverage for plant spacing and safe distances from plants for placement of fixtures.

UNITS OF MEASUREMENT

Marijuana growers are always looking for ways to provide optimal conditions for harvesting premium crops, and measuring light intensity is a method of determining needed improvements for achieving the best light. Both the lighting industry and horticulturists rely on light values that include foot-candles. A foot-candle is a measurement of light intensity; best described as a one square foot surface with a uniform distribution of one lumen of light (approximately 10.764 Lux). Lux, abbreviated lx, is equal to one lumen per square meter. You will not be using these units of light measurement often, but they are useful when designing your system, inspecting your present system, or when selecting bulbs.

OPTIMAL LIGHT FOR CANNABIS PLANTS
Ideal ranges to maximize growth and flowering (applied to leaves)

PHASE	FOOT CANDLES	LUX	HOURS PER DAY
Seedling, Clone	372-465	4,000-5,000	16-24
Vegetative	2,323-4,645	25,00-50,000	16-18
Bloom	4,180-6,503	45,000-70,000	12

LIGHT OBJECTIVES

As a basis for comparison, full sunlight has an intensity of 10,000 foot-candles, but be aware that optimal ranges based on a plant age and stage of development change over the life of the plant, and full, direct sunlight illuminance is not the goal through the growth stages. Just as you would not place a marijuana seedling in a full sun or for an extensive time for best results, a mature cannabis plant in full flower benefits from bright light, but for 12 hours with 12 hours of darkness in an artificial environment.

For outside growers, cannabis hybrids will grow in shady conditions but the flowering will be minimal, if at all, and has no significance to a commercial grower of premium cannabis. It is important to cultivate your plants with a minimum of 6 hours of sunlight per day; more is better. Outside growers benefit tremendously by relying on the natural seasons and plan their crops; accordingly. The first day of spring, the vernal equinox (around March 20 in the Northern Hemisphere) is a good time to start or prepare to start a cultivation project outside. Summer is a good "midpoint" for a cultivation project when relying on natural lighting. You may start a grow cycle after the summer solstice (around June 21 in the Northern Hemisphere), but your results will be much better if it is before to allow enough vegetative time, about 4 weeks and about 8 weeks in the bloom phase before the weather changes in the fall months.

Whenever possible, offer your plants uniform and penetrating light. Providing even lighting, directly above plants helps to avoid phototropism and will lead to balanced branching and growth near the top and on the outer tips as the plant "reaches" for the source. Directing light uniformly on the sides then helps to promote even and balanced growth, although growers rely on reflected light for this task as overhead fixtures are more practical and therefore widely used.

To help ensure the best light penetration to the inside and lower portions of a plant, groom out large leaves regularly and remove small branches, especially from the inside area of the plant (near the trunk) that will not produce significant flower buds. The procedure may need repeating several times in the vegetative phase as this pruning action usually results in new leaves or branches at the cut site.

An added benefit of proper lighting is the discouragement of some kinds of pests and pathogens that thrive in dark, damp conditions. Plants located in corners of indoor facilities often produce less blooms on the darker sides of the plants, offering a haven to pests who feed on the lush new growth that occurs in shade. Simply adding light to all parts of the growing area helps resolve both issues.

In summary, light will have a tremendous effect on the quality and weight of your crop. If you are an outdoor grower, provide as much sunlight to your plants as you can, spacing them to prevent shadows. If you are an indoor grower, provide light in the right spectrum for vegetative growth and then the right spectrum for the bloom stages of maturity. Controlling the duration and intensity are factors that offer precise requirements to rapid, hardy growth and bloom for indoor or tent growing; an advantage if you know how to manipulate it like an artist in specialized agriculture.

"Light Dep" is a shortened version of light deprivation, a technique to induce flowering. In cannabis, light reaches plants only 12 hours per day while the other 12 hours is in total darkness. For greenhouse growers, kits are available to block out daylight for this method. Within an indoor facility, timers or controls restrict light to begin the bloom phases of development and finish the harvest precisely according to schedule.

A male cannabis plant thrives in the sunlight and open air of a natural environment.

Replicating Nature

In an artificial environment like in a greenhouse or indoor cultivation site, we attempt to reproduce what nature does for the benefit of our plants. The objective behind all new climate control technology, including lighting, is to identify what cannabis enjoys and then maximize it to benefit propagation, cultivation, and crop production.

There are many priorities in cannabis cultivation and so everything has a rank of importance to a serious grower who must work within the confines of space and funds. When it comes to lighting, the good news is that there is a system for nearly every garden and budget. The point is that the first step is to provide the best light for your plants according to their stage of development and secondly, to make it work for your specific operation; both physically and financially.

Enhanced lighting also does not need to be an all or nothing proposition. Any improvement to an environment and the climate used for cultivation is good and while it may be costly to have the latest and greatest power consuming machines ever made, it really is unnecessary. The goal is to get close to targets of optimal conditions, not go overboard with equipment that may be costing more to purchase and operate, than it is worth.

Most lighting systems are long-lived with bulb life measured in the thousands of hours. Careful selection applicable for your needs before the investment and installation are a practical way to save money, provides benefits to the crop, and you end up a smarter cannabis farmer as you get closer to matching nature's ways.

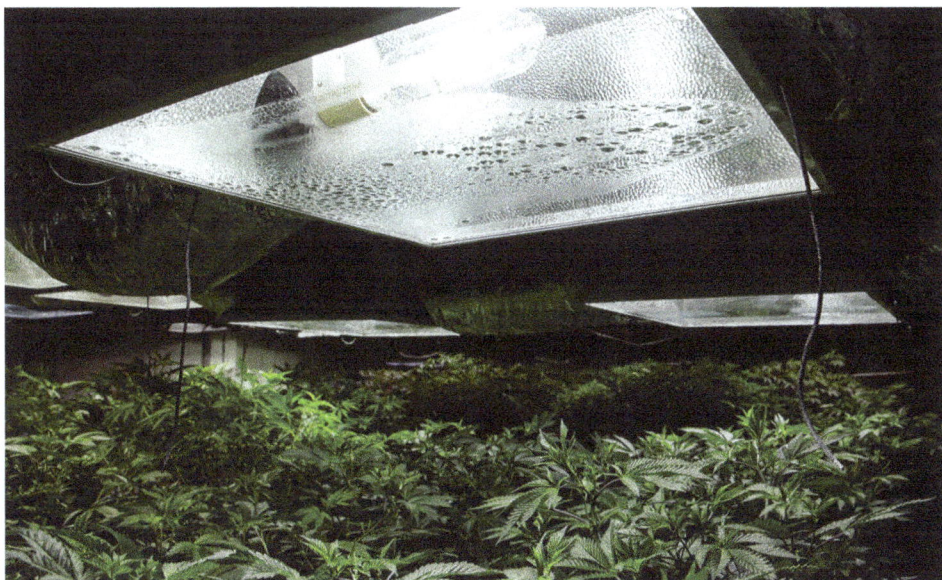

When selecting a lighting system for new installation or retrofitting for efficiency, do not jump right in and purchase what you may not want or need. Instead, determine what type is best for your budget, grow space, and which will give your plants the best light for maximum crop production. Then, talk to some experts, find a good electrician, and shop around; the effort will pay off with premium production that is profitable and a facility that is safe.

It is uncertain how much marijuana cultivation occurs outdoors, in greenhouses or within enclosed buildings. Farmers may have preferences based on a few factors such as production goals, schedules, weight, taste, aroma, and even marketability, but growing premium cannabis is possible in all locations.

As climates vary widely in areas where cultivation is lawful, many growers are starting or moving in an indoor direction for all or part of their growing. It provides cultivation opportunities year-round, protection for the plants, but more important, it affords total control over environmental factors like light.

The gaining popularity and practicality of indoor cultivation have resulted in improved technologies for better plant growth and yield. Energy efficiency continues to improve right along with a huge selection of bulbs, fixtures, hoods, vents, and automation. When choosing a light system, calculate not only the initial cost for startup, but also the cost to operate the system from seedling to harvest, since utility costs have a large effect on crop production costs.

•Fluorescent

Fluorescent grow lighting for cultivating cannabis is an economical way to provide artificial light with minimal heat production and low energy use.

T-5 (Tubes)

Compact (CFLs)

-Pros

Inexpensive to purchase, economical to operate, minimal heat production, good light spectrum, useful in small or tight spaces.

-Cons

Best for small clones or plants in vegetative stage as crop yields are lower when used for bloom cycle, poor light penetration below plant canopy.

•High Intensity Discharge (HID)

Light emits from bulbs screwed into a hooded receptacle, often powered with a fan exhaust that allows heat discharge since the bulbs get very hot.

High Pressure Sodium (HPS)

Metal Halide (MH) (Available in daylight or blue spectral ranges.)

Ceramic Metal Halide (CMH)

-Pros

Most commonly used in commercial gardens, provide high yields, economical to purchase, easy to install.

-Cons

Produce high heat, require special socket, high cost to operate.

•Light-Emitting Ceramic (LEC)

The reduced heat output, reduced energy use and an optimal spectral range makes LEC a good choice for indoor or small spaces and lamps have a typical long-life.

-Pros

Considered the most efficient grow light, easy to use with minimal height adjustment from plants, produce high yields.

-Cons

High heat production requiring exhaust, most setups require venting to exterior, multi-part fixtures requiring bulb and external ballast with cabling.

•Light Emitting Diodes (LED)

Popular for cannabis growers as an alternative to HPS lighting. LED lights run cooler and often have a fan built into the fixture.

-Pros

They run with less heat than HID, plug directly into the wall or ceiling, heat discharges away from plants, good light penetration to lower parts of the plant. Some growers report a more resinous flower bud by using LED alone or with HPS; however, the hard numbers are not currently available.

-Cons

Higher watt LED fixtures may require external venting for heat exhaust, smaller yields than HPS grow lights.

SYSTEM COMPARISONS

This list is a random sampling of products for the types of lights designed for growing. Prices in effect at time of publication from various sources; subject to change, with no guarantee of availability. Brand names and trademarks belong to their respective owners.

CFL: Hydrofarm® Fluoring CFL, 6400K $75.95

FLUORESCENT: Sun System® Sun Blaze T5 HO 24 Fluorescent; 4 lamps, 24" fixture, Blue 6500K MSRP: $123.95, Discounted $110.35 (Red spectrum lamps available for bloom)

LEC: Sun System® 315 Watt LEC Fixture; 3,100K LAMP MSRP: $609.95 Discounted: $453.96

HPS/MH: Solis Tek® 600W Air-Cooled HPS/MH Full Spectrum Grow Light Kit; includes air cooled reflector, MH Lamp, HPS lamp, ballast Estimated price: $325.00 to $475.00

LED: California Lightworks® SolarFlare 220W Grow Light, Spectral Blend: Full Cycle MSRP: $519.00, Discounted: $439.00

•Timers and Controls

Standard Timers: A wide selection includes 24-hour timers, 7-day timers, multi-outlet timers, timers with heat sensing thermostats, power strips with timers, digital and analog timers.

Power Boxes: Distribution boxes designed to run more lights with fewer ballasts. Use with limited power circuits; professional advice recommended.

Light Controllers: Bluetooth® and phone apps are available to control lighting digitally and remotely.

Photocell Controllers: Day/Night sensors control, On/Off times for lights and fans plugged into the device.

•Grow Light Options

Movers: Systems use rails suspended above garden to facilitate movement of light fixtures. Automated movers utilize a motor driven pulley to move lights on a timer. Kits are available that start in the $500.00 range.

Hangers: Various methods utilized to hang lights, fixtures, or electrical cords. Hooks must be secure, anchored safely into wood or metal for any type of hanger used; the last thing you want is for a device to fall on someone or your plants. Very popular for safety and convenience is a ratchet and pulley system with nylon rope to raise and lower devices, once attached to an eye hook or secure fastener.

LIGHTING ACCESSORIES

Grow light systems often require accessories beyond the lamp and fixture for maximum control and efficiency. For instance, reflectors and hoods that are suitable for the type of bulb are usually an add-on or upgrade item, while venting and exhaust fans for light produced heat are also items for separate purchase.

Hoods: Reflective devices over light fixtures increase the intensity of light to plants and influence the size of the growing space by uniformly directing it to the plants. Higher yields are the goal with the use of hoods.

Vents: Heat produced by grow lights, especially HID types, exhausted from the grow area or tent by hood vents with fans and ducting to an external destination.

Heat Shields: Attaches onto fixtures and hoods to direct or contain heat for better control.

Light Meters: Most useful to determine the luminance (foot-candles). A good quality digital meter runs about $50.00.

Reflective film: Once very widely used, growers now rely more on highly reflective white paint for walls and ceilings, chiefly for ease of cleaning and sanitation. The special film weighs little and is helpful in certain settings; however.

Lighting Terms for Growers
useful definitions for planning systems or purchasing supplies

Amp: A measurement of the intensity or strength of electric flow.

Arc: Luminous discharge of electricity between diodes in HID lighting.

Average Rated Life: Life term assigned to a specific lamp type.

Ballast: Electrical device to start and control flow of power to gas discharge lights.

Base: A mechanical device that connects a lamp to an electrical source.

BU: A code to indicate the bulb must operate in a base up position.

Candlepower: Luminous intensity of a light source.

Cold Start Time: Length of time required to bring a HID light to 90% light output from a cold position.

Color Temperature (Kelvin Temperature): The unit of measurement to define the spectrum of light from a lamp.

Cost of light: Fixture, lamp, installation, maintenance labor, and power costs.

CRI: Color Rendering Index is an international system to rate a lamp in rendering color. (The difference among lamps is usually insignificant).

Fixture: Electrical fitting designed to hold components of a lighting system.

Fluorescent Lamp: Phosphor coating transforms ultraviolet energy into visible light. Useful for seedlings and clones.

Foot Candle: Measurement of light intensity; equal to the light from one candle at a one foot distance.

Halogen: High pressure incandescent lamps containing halogen gases. Not effective for grow lighting use because of limited spectral range.

HID: High Intensity Discharge lamps are a group containing Metal Halide, High Pressure Sodium, and Mercury.

High Pressure Sodium: (HPS) Lamps ignite sodium, mercury, and xenon gases within a sealed tube.

Hood: The reflective cover for a HID lamp.

HOR: Code indicating bulb must operate in a horizontal position.

Incandescent Lamp: Thin wire heats to produce light. Not effective for grow lighting and produce excessive heat for horticultural applications.

Illuminance: The density of luminous flux on a surface.

Intensity: The magnitude of light energy per unit, diminishing further from the source.

Kelvin Temperature: Unit of measurement of color from a lamp.

Kilowatt: A unit of electrical power; equals 1,000 watts.

Lamp: A bulb or tube charged electrically for light.

Light Mover: A motorized device that moves HID lighting back and forth above cultivated plants.

Lumen: A measurement of light output; equals one candle on one square foot of surface located one foot away.

Lux: A unit of illuminance; equals one lumen per square meter.

Metal Halide Lamp: Light produced from arcing electricity in metal halides. Produces a blue-white spectrum useful in vegetative growth.

Mercury Vapor Lamps: A HID lamp that produces light from arcing electricity in mercury vapor.

Mogul Base: A large Edison screw base for lamps.

Photoperiod: The expression of hours of light within a twenty-four-hour period.

Photosynthesis: A growth process of plants producing carbohydrates from light energy, water, and CO2.

Phototropism: The movement of a plant toward a light source.

Spectrum: A linear expression of electromagnetic wavelengths.

UL: Underwriters Laboratories Inc., is an independent product testing and certification organization.

UV: Ultraviolet Light within a wavelength of 100-400nm.

Watt: A unit of power that measures the rate of doing work; in lighting, it is the measurement of the radiant luminous-flux of a lamp.

Safety Specs

Exposure to constant sunlight and High-Pressure Sodium (HPS) indoor lighting is a concern of the cannabis farmer for eye health and safety, but also for the ability to see properly under intense light. Fortunately, manufacturers saw a need by growers and developed specialized lenses for protective eyewear.

Costing in the range of about $20.00 to $25.00, glasses are available to reduce glare, especially in the red to orange spectrum and provide both UV and IR protection as well. This is a distinct advantage since mites and other tiny pests are difficult to see with light glare in this warm spectrum, causing some growers to turn off the HPS lights just to inspect their plants if they do not have anti-glare glasses.

Some eyewear styles offer interchangeable lenses for both indoor and outdoor use, while others sell separately based on use location. Most eyewear for grow rooms is of durable plastics that are ideal for the workplace and very easy to keep clean. These specialized glasses are available from most hydroponic stores and suppliers, but also inquire with eye care professionals for application of prescriptive lenses that are applicable for protection from bright light exposure.

Eye protection is important for not only ourselves and our employees, but also for visitors to our operations, so keep a few pairs available for them, new hires and folks who always seem to misplace their own.

Chapter 8
Air: The Invisible Influence on Plants

Air quality has a direct effect on plant growth. Outdoor growers have some influence over air protocols for cultivation simply by site selection; however, greenhouse and especially indoor growers have measured control over air quality and conditions. This is a substantial advantage to the premium bud grower who can provide an ideal growing environment based on each stage of growth.

Air and systems of filtration, sterilization, and ventilation are of interest to indoor growers who use various defined spaces from grow tents to warehouse gardens. Fresh, clean air is critical at the proper temperature and humidity for plants to flourish and reach peak performance in production.

Expert growers often take cultivation to new heights and air management for cultivation is no exception. The addition of CO_2 into the grow room is a method of increasing growth rates and bud formation to higher weights and qualities. This Chapter finishes with a look at adding CO_2 to your cultivation space as an option for improving air quality for your plants.

8 | a. Optimizing the Environment for Crop Production
Air Quality and Harvest Bounty

Air quality for cannabis plants is really about the best climate you can provide for their health and vigor. That includes everything about the air; the movement, the content, the temperature, and the humidity. By focusing on the combination of factors in total, managing an ideal environment for maximum production becomes an easier task.

For a myriad of reasons, marijuana cultivation utilizing hydroponic culture methods is on the increase, affording opportunities for complete environmental control and optimization. The atmosphere of the indoor growing space plays a significant role in the health of your plants and the quality of your harvest. With solutions for almost every conceivable cultivation project that occurs indoors or in greenhouses, improving the air for your marijuana plants has never been easier or more affordable with results that are efficient while boosting production.

Outdoor growers should also be aware of air standards that are helpful to healthy and vigorous plants since starting new plants indoors or in a greenhouse when they are vulnerable or need warmth from cold temperatures is a great way to get a jump on the growing season. Just a few extra weeks of protection will lead to larger plants that are hardy and ready to take on the great outdoors.

As energy costs continue to increase with greater effects on the environment, it is wise to plan for efficiency when starting and sometimes it is even smarter to retrofit old setups to achieve budget and ecological goals in utility consumption. It is possible to grow premium cannabis within the safe ranges of temperature, humidity, and CO_2 level, but your chances improve substantially the closer you maintain the ideals of each.

ELEMENTS OF ARTIFICIAL CLIMATE CONTROL:

- Temperature
- Humidity
- Movement
- CO_2 Level
- Odor Control
- Fresh Air Filtration
- Stale Air Exhaust

Examining the options in controlling the atmosphere in your growing area or facility is an important first step to maintaining the correct climate for your plants at various stages of development. Precision is the key, so choose the most accurate devices and appliances that you can afford, ranging from all-in-one controllers to individual controls of heating, cooling, humidity, CO_2 delivery (if used) air movement and ventilation for fresh air entrance and stale air exhaust.

New wireless technologies offer the ultimate control system for climate, including air movement, intake, and exhaust. Combining "smart" plugs, timers, and controllers can create a system that provides not only a way of monitoring, but adjusting to optimum levels also.

Of the various climate control devices and appliances on the market, many have built in thermostats or regulators for control. Some have accuracy ranges that vary, so climate monitoring and measuring instruments that are separate and independently located are an effective way of maintaining the precision you are after. It is not overkill to have multiple thermometers or hygrometers in different locations within the growing environment.

AIR MANAGEMENT OBJECTIVES

To obtain the most desirable climatic conditions in your quest for premium grade cannabis, consider this path of air flow as the goals and priorities of creating an ideal environment for their growth and flower development.

Fresh Air > Filtered > Warmed/Cooled > Dehumidified/Humidified > CO2 injected > Moved through plants > Deodorized > Exhausted

When you have developed a general plan for climate control, get specific with a shopping list that identifies the power consumption of your climate control equipment. The next section, Climate Control Equipment, describes in greater detail the choices that are available. Make certain that along with your lighting energy demands you have an adequate power supply for total watts and amps that you also need to operate your cultivation tent, room, greenhouse, or facility. If you are uncertain, contact an electrician. Electricity is nothing to fool around with if you are unqualified. Hazards including shock and fire can result from faulty or inadequate electrical wiring. Consulting a licensed and qualified electrician is money well spent before you rush out and purchase tons of equipment or place you or someone else at risk, not to mention your property and plants.

Not all growers will elect to setup total systems for air management for various reasons, most likely expenses, or budgetary limitations. In those cases, it is important to at least keep the air moving around the plants to promote health, vigor, and discourage conditions that create disease like mold or fungus. For healthy plants, temperature and humidity ranges are also key factors for growing cannabis plants in artificial settings.

CLIMATE OBJECTIVES

STAGE	TEMPERATURE	HUMIDITY rH (Day)*
Seedling, Clone	Day: 82°, Night: 70°	75%
Vegetative	Day: 80°, Night: 70°	55%
Pre-flower	Day: 78°, Night: 65°	50%
Bloom	Day: 74°, Night: 60°	40%
Harvest	Day: 65°, Night: 60°	20%

Night humidity ranges about 10% below day time targets.

DEODORIZING YOUR EXHAUST

As a cannabis grower, it is always wise to keep your neighbors happy, residentially, or commercially. Odors that you may enjoy may be offensive to others who may complain to the authorities or licensing agencies. Certain cannabis odors may also attract the attention of thieves looking for free pot.

There are a couple of methods of deodorizing the air in your growing space or the exhaust air from facilities like indoor cultivation gardens, tents, or greenhouses. Exhaust filters are available in popular sizes to fit most ducting, from 4" to 12". Another approach uses sprays, gels, and mists with neutralizing scents dispersed from refillable dispensers, with models to fit all budgets.

Keep in mind that good planning will save money eventually. While you do not need to purchase the latest and greatest technologies for a bountiful harvest or purchase the most expensive solutions, your chief concern should be creating the most ideal growing conditions that you can. The air surrounding your marijuana plants is a huge influence on growing premium grade flowers. With the knowledge, some planning, and a marketplace full of choices, you can make that air near perfect.

Move That Air

Air movement is a central part of establishing an artificial climate that is advantageous to plant health. It may not be ideal in every way, but at minimum, the air in your tent, grow room, or greenhouse needs to have a constant flow through the canopies of your plants. It is an inexpensive way to reduce stagnant air, cool the plants slightly and help plants establish strong structures, in branches, leaf formations, and sturdy root systems.

Plants gain most of the carbon dioxide and other essential gases they require through the stomata on the leaves. The tiny pores must remain free from dust or water that would preclude necessary air exchange for the plants to survive. For the commercial grower, this small aspect is an important part of morphology of the plant and the premise behind basic but beneficial cultivation methods; keep the air fresh, clean, moving, and the leaves free from dust and water.

Retailers and dispensaries that sell clones often use small, clip-on fans to gently move air through the young cuttings. It helps primarily by reducing over damp conditions and assisting in cellular strengthening of tissues. For propagators, they are great in that they do not blow the plants away but rather provide just enough air to accomplish the needed tasks. They are also very convenient for use on wire shelving, also widely used for good air circulation. Most fans for horticultural use have adjustable fan speeds, and even the small clip-on type includes this feature.

The wall mounted fans sixteen (16) inches in diameter and larger have adjustable speeds, tilting options, and oscillate; very convenient for indoor growing. Pedestal and box fans have uses also, but may take valuable floor space. As plants mature, they will block air flow patterns that existed before they got larger, so multiple fans are usually the best solution.

Fans cost about $25.00 for the small 6-inch clip-on type (avoid inexpensive units; they do not last for more than a year), with wall mounted units starting under $100.00. With a goal of moving as much air volume as possible in all areas of the grow space, plan for locations that are near electrical outlets and avoid extension cords. Fans are excellent pieces of equipment for indoor growers and when used in combination with other air systems, a good investment in the pursuit of premium cannabis.

8 | b. Climate Control Equipment
Costs and Benefits to Crop Yields

Any size indoor garden, from a tent to commercial warehouse, will require equipment to monitor and modify the climate to create conditions best suited for plant growth. The premium crop farmer can make the indoor weather conditions like heaven to cannabis plants. Remain mindful of the goals and design an air management system that will be energy efficient and give your plants what they want and need on the way to producing a bumper crop of desirable character.

A very cool (no pun intended) part of cannabis cultivation is the emerging technologies and products to support growing the best weed you can imagine. It can be expensive; however, so to avoid costly mistakes, consider that purchasing the best you can afford is never a bad idea, but the best is not always the most expensive. Search for items you might purchase in the market, online and at brick, and mortar stores like hydroponic supply shops and plant nurseries; see what is available before spending a dime.

Unless you have unlimited funds, prioritizing what is essential for optimal atmosphere conditions for your cannabis plants is a great way to save some money, get what you need and watch your plants flourish. The life of most appliances last a while, so invest wisely and enjoy the benefits.

This section of the book contains random samplings of products representing various aspects of controlling climate for growing within an interior space. The listed items should provide the reader with an overall view of equipment types and uses. Prices in effect at time of publication from various sources; subject to change, with no guarantee of availability. Brand names and trademarks belong to their respective owners.

FRESH AND CLEAN AIR

The importation of fresh air into an enclosed greenhouse or indoor growing space is vital for plant health, but it is important for it to be clean air. Fit air intake ports with a filter to prevent entrance of pathogens or vectors like dust, mold, fungus, insects, arachnids, rodents, etc. Carbon and HEPA filters provide the best protection. Some are disposable with described life-spans while others make cleaning and sanitizing an option for reuse.

Ducting or entry ports can be a square but are usually round starting with 4" diameters that go as high as 12" with readily available filters to accommodate 4", 6", 8", 10" and 12" vents. Mounting or attachment is easy with clamps or standard fittings often included with the filter.

Each filter type and size requires adequate clearance around the unit for best results. Most suggest installation lower on walls instead of above plant canopies where air exhaust should take place. This provides temperature benefits (cool air falls, warm air rises) and supports a healthy air flow around your growing plants, where fresh and filtered air replaces stale and spent air or air that is too warm.

AIR CLEANING

-Horti Control™ Dust Shroom HEPA Filter 6" $119.00

-Phresh™ Intake Filter, 6" x 12" $84.95

-International Growers Supply™ Organic Air 6" HEPA Air Filter $139.99

(*Sterilization equipment for air systems is also available.*)

TEMPERATURE CONTROL

Cannabis plants do not produce well when exposed to temperature extremes, even for short periods of time. Maintaining a safe temperature in your cultivation atmosphere is important if you expect a decent crop; imperative if you want premium grade flowers.

In most indoor growing situations within temperate climates, it is more likely that you will need to cool the environment instead of warm it, especially when using certain growing lights. Heat produced from HID growing lights can easily cook your plants if the heat has no exhaust or not cooled by air conditioning. It may also be necessary to provide cooling to nutrient solutions using chillers rated for solution used or stored in reservoirs. Improper solution temperatures can harm root systems and influence air temperatures in cultivation spaces.

In colder regions or in winter months, it may be necessary to heat the indoor growing environment or warm nutrient solutions to proper temperatures for optimal results. In both heating or cooling, provide a source that will not blow heated or cooled air directly at plant foliage as this will dehydrate marijuana plants very quickly. This is especially the key when selecting heaters.

Within the realm of climate control are numerous thermostats and controllers that offer options to tailor any size garden. Although some are pricey, the reliability of automated climate control provides a degree of protection that often pays for itself over time; both in labor hours saved and in peace of mind.

• Heating

Most suppliers of hydroponic equipment do not stock heaters, except for those that warm water tanks or nutrient reservoirs. The general hardware marketplace has a big selection of all types of heaters that may be suitable for indoor plant cultivation and so it is just not practical to stock a full range of heating options for specialty retailers in horticulture. When choosing the right one for your cultivation efforts, select a heating system that is energy efficient and rated for the space you need to heat; a cubic foot measurement of (length) X (width) X (height), or a square foot measurement depending on the one used by the manufacturer for suggested heating space size.

Forced air heating systems, structurally attached, rely on oil or gas; radiant systems at baseboards or in floors heated by electricity, oil, or hot water. Except in large commercial gardens, portable heaters are usually more practical and more affordable and the focus here instead of hard wired systems that require permits and professional installation.

Space heaters, also called portable heaters, get power from electricity, kerosene, natural gas, or propane, and include ceramic, infrared, and oil filled. Within those types include models with thermostats, timers, fans, angle adjustments, or oscillation ability.

> BTU, British Thermal Unit is a unit of heat needed to raise one pound of water by one degree Fahrenheit. 1 BTU/hour = 0.293 watt.

Determine the heating factor for your climate zone, typically available from your utility company. This number provides an estimated BTUs required per square feet to heat or cool an insulated structure in your area. Simply multiply your square footage by this BTU/sq. ft. to determine the BTU requirement for your setup. For uninsulated structures or greenhouses, consult the recommendations of the heater manufacturer or supplier for space measurements that the unit will heat or cool efficiently and safely. Also, it is common for retailers to list or group heaters by the room size that they will efficiently heat.

Propane

Look for models that include adjustable heat settings and electric ignition.

- Dyna-Glo® Pro 60K BTU Forced Air List $100.00

- ProCom® 60K BTU Liquid Propane Forced Air List $120.00 (Heats about 1,400 sq. ft., runs up to 22 hours on a single 20 lb. LP Tank.)

Kerosene (Non-vented)

Best utilized as an emergency source of heat (or light).

- DuraHeat® 23,000 BTU Portable Kerosene Heater (Heats up to 1,000 sq. ft.) List: $140.00

- Dyna-Glo® Pro 80K Forced Air Kerosene Portable Heater (Heats up to 1,900 sq. ft.) List: $200.00

Electric (Basic convection)

Ideal for small space heating.

- Patton® 1,500 Watt Recirculating Portable Utility Heater (Heats up to 200 sq. ft.) List: $40.00

- Lasko® 1,500 Watt Low Profile Silent Room Heater (Heats up to 300 sq. ft.) List $54.00

Ceramic

Provides fast heating, but only useful for small rooms or spaces.

- Lasko® 23 in. 1,500 Watt Digital Ceramic Heater with remote List: $58.00

- Stanley® 1,500 Watt Utility Ceramic Heater with Pivot List: $50.00

Infrared

Spreads even distribution of warmth for single room heating without dehydrating.

- Lifesmart® Large Room 1,500 Watt 4 Element Infrared Heater with Remote (includes washable air filter) List: $60.00

- Duraflame® Dartmouth 1,500 Watt 6 Element Infrared Quartz Heater (Stays cool to the touch, heats up to 1,000 sq. ft.) List: $115.00

Oil Filled

Provides radiant heating for small rooms; very portable.

- Pelonis® 1,500 Watt Digital Oil-Filled Radiant Heater, heats 100 sq. ft. List: $60.00

- DeLonghi® Full Room Oil-Filled Radiant Heater, heats 144 sq. ft. List $95.00

• Cooling

Air conditioning is an aspect of indoor farming that is usually not an option. The heat generated from cultivation lights, pumps, fans, etc., require use of some form of air conditioning to keep temperatures within optimal ranges. Fortunately, there are several choices to fit almost every budget.

First determine the BTU requirement for your garden area. For appliances that provide air conditioning, the BTU measures the heat that an appliance can remove from a room per hour. Technology using refrigerant and water removal from the air combines in various models to accomplish air cooling. Some need hot air discharge to outside locations, others use evaporative condensation removal.

While variables including building type and insulation value influence cooling, use this guide to help in estimating your BTU requirement:

Room Size (sq. ft.)	Suggested BTUs
150	5,000
250	6,000
350	8,000
450	10,000
550	12,000
1,000	18,000
1,600	25,000

TYPES OF AIR CONDITIONERS FOR GROWERS

Permanently installed systems are the ideal, followed by window or through the wall types. Marijuana growers also use three other types; all with unique attributes and readily available from several sources including online and your local dealers.

Portable

Hydroponic equipment dealers often stock portable units in the $500 to $1,500 range, but most will order whatever size a unit that you require to properly, and efficiently, cool the growing atmosphere.

- Ideal Air® Dual Hose Air Conditioner 12,000 BTU. Condensate water used in the cooling process expelled as a fine mist. List: $500.00

- Active Air® 14,000 BTU Portable Digital Air Conditioner. Uses window kit for air exchange. Popular choice for many small growers. List $600.00

Mini-Split Systems

Efficient and dependable, this type of system has an indoor unit and an outdoor unit, connected by flexible copper lines. Most models require professional installation by HVAC professional.

- Aura Systems® 12,000 BTU Mini-Split Air Conditioner. Uses quick-connect fittings for easy installation. List: $1,200.00

Ceiling Mounted

Designed to handle large volumes of air by recirculation without external ducting.

- Surna® Ceiling Mounted 2-Ton Air Handler. List: $1,600.00

HUMIDITY CONTROL

Keeping humidity ranges within an optimal range (varies during growth cycles) is essential for plant health. Over damp conditions can harbor diseases that are very harmful to cannabis plants and most strains like a drier atmosphere; like a temperate climate zone. Also, ideal night humidity levels are slightly lower than daytime.

Note that maintaining proper humidity levels can be a challenge within a sealed, indoor, growing environment so carefully select equipment that will do the job for the space you have. A constant moisture buildup occurs in any size growing room from irrigation, evaporation of the growing medium and from plant transpiration. Rarely are natural humidity levels too low when growing indoors, so utilize dehumidification appliances for successful cultivation.

- Active Air® Dehumidifier - Analog Controls, 43 Pints Per Day of water removal. List: $275.00

- Ideal Air® Commercial Grade Dehumidifier 60 Pint. List $595.00

MOVEMENT

Good air circulation around your plants serves many purposes. The stomata on the marijuana leaves gets ample access to fresh air while moisture buildup keeps to a minimum when fans move air through your garden. Foliage is cooler, and plants develop better supporting structures when exposed to mild breezes or gentle air flow. Blowing air directly at a plant may cause harm by accelerating dehydration, so tilt fans or adjust oscillation as required.

The options available allow growers to fine-tune vital air movement with various types of fans and blowers, including speed controls for these popular types:

•Wall mounted Fans

•Clip-on Fans

•Pedestal/Floor Fans

•Inline Fans (for ducting)

- Active Air® 16" Wall Mount Fan. Easy mounting and a popular choice for many growers. List: $ 60.00

- Hurricane® Super 8 Digital 16" Wall Mount Fan. This popular unit includes a remote control. List: $76.00

EXHAUST

Just as important as fresh air entering your grow space is the exhausting of stale air. A ceiling or wall mounted exhausting fan is the best solution, requiring no specialized horticultural equipment. A variety of parts including ducting, filters, flanges and fittings, clamps, reducers, etc., are available to complete any exhaust system with some simple planning. Advice and installation from a professional are useful and may necessary.

A complete exhaust system should include an exhaust filter to deodorize any smells that may be offensive to others. Locate exhaust fixtures away from property lines or from areas where the noise of air exhaust might inadvertently attract attention. Alternately or with an exhaust filter, use dispensers, gels, sprays, and mists, listed in sample pricing.

- Covert® Carbon Filter, 4" X 12", 200 Cubic feet per minute. List: $90.00

- Ona® Cyclone Fan Dispenser, 121 CFM (controls about 5,000 square feet) List $140.00 Place over a five (5) gallon pail of Ona® gel, List $189.00

CO2 ENRICHMENT

Enclosed environments must import fresh air to keep carbon dioxide levels applicable for plant growth. The natural atmosphere contains about 300 parts per million (ppm) of CO_2, but if levels within your growing space drop below 200 ppm, plant growth ceases. Rapidly growing marijuana plants will consume increasing amounts of CO_2 as they develop and may deplete safe levels if not monitored. However, increasing the CO_2 levels within the artificial atmosphere can substantially increase vegetative growth and bloom development in cannabis, resulting in many commercial growers moving to CO_2 rich environments.

A target for many high-end indoor growers who use CO_2 enrichment is 1,500 ppm. Achieving that CO_2 content is practical with controllers, monitors, regulators, tanks, and generators. Carbon dioxide enrichment offers many options for installation and uses. To complete a system that is suitable for your space, requires the purchase of separate components that are readily available from various sources.

Any CO_2 system requires a controller/monitor, dispersing hoses and a regulator that connects to the source of CO_2; either from a generator that produces it from burners or from tanks usually exchanged for full tanks at retail stores offering CO_2. Keep in mind that generally you will be receiving used tanks in exchange for the empty tanks you turn in for a fee that includes the tank rental and the CO_2 in it. CO_2 filling services are hard to find, so this system of tank exchange is very common. Hydroponic dealers typically offer CO2 equipment, and gas tank services; check them first.

Some small growers using tents or small spaces may opt for CO2 bags or boxes that produce continuous CO2 supplies from biological activity within the container. This economical approach requires no control, monitoring, or regulation equipment. Purchase enough bags or boxes of the product to meet your cubic foot space demands for the most effective use.

There are many variations of useful CO2 systems, but they are site-specific to your individual setup, so no model comparisons appear here; shop methodically and compare features along with prices. The total cubic feet of your growing space will be useful in determining what fits your budget and what are the best supplies and equipment for your layout. Most carbon dioxide systems rely on that measurement for proper set up. Dealers specializing in CO2 systems will help in planning and design and fortunately most entry-level equipment is affordable, while also expandable. If you seek the latest and greatest in cannabis cultivation, consider CO2 enhancement a good investment for the benefit it provides.

CO2 generators create and maintain levels when paired with a controller unit and a viable solution for enrichment. They require a gas source, either Liquid Propane (LP) or Natural Gas (NG) along with electricity for the generator and controller. A two (2) burner unit producing 4,500 to 5,500 BTU/hr. (depending on the gas source type) is a good size for a small propagation room of about 100-120 square feet. Typical pricing is about $300.00 for the generator of this size and about $500.00 for a digital controller. Additional assembly or equipment parts are usually additional, but for the long-term grower, a total generator package suitable for a specific growing space is worth consideration.

Grades and Standards

The term "premium" has different meanings to different people. In the cannabis world, it is a subjective term since there are no universal grades or standards of quality yet determined for marijuana. For most in the marijuana business; however, premium stands for an expectation of physical characteristics that set it apart from ordinary. More important, the "high" or benefit from the flower or an extraction product from it, is the supreme factor in rating the quality of cannabis and judging it premium.

What gives a trimmed bud a top rating? First, it must be hard, dense, and full of dried trichomes, or "crystals". No seeds, please and stems must be almost invisible. Stickiness is a plus, although folks should not handle the merchandise; buyers or sellers. The color of buds is influential in that it gives a product "bag appeal." Consumers look for bright greens, orange hairs, and anything purple.

Next up is the smell, aroma, or "nose" of the finished product. Some consumers prefer a stinky, skunk odor, while others might select one with a fruity essence. Still, others want a hard diesel pungency to their herb. Identifying characteristics that determines value will help you select those strains to grow.

Consumers read labels, and testing results play a big role in a quality rating, albeit a personal standard, for purchase of your product. How and where the cultivation occurred has a direct effect on purchasing decisions. Some prefer sun-grown flowers, while others are looking for chemical values in the finished herb and do not care if it came from grandma's basement. While appearance is important and terpenes of the flower help define the individual aromatic attributes, most consumers ultimately decide value based on the THC or CBD content or the expected effects from consumption, a sobering reality to growers who may have an unrealistic view of the quality of their crop. So, while the goal is a premium cannabis crop and the subject of this book, an objective evaluation with a lab test is ultimately what determines the value of your crop in the legal marketplace.

Market forces from supply and demand affects the price of any cannabis commodity, but value is a bigger factor, judged by the quality of the product. Generally, if there is a flood in the marketplace with poor quality flowers, the price will go down because no one wants it at any price. Alternatively, a good quality crop will hold value better and command a higher price because cannabis has a long shelf life and eventually demand will catch up with a limited supply. At the end of the day, lacking any universal grading or standards, growers must always strive for a superior product; anyway the consumer measures it.

Chapter 9
Harvest, Cure, and Store: Finishing the Crop

Thriving cultivators of any prized commodity understand that it is the final steps of production that has a substantial effect on the ultimate quality and value of a crop. Considering all the work and expense that go into marijuana cultivation, the harvest, preparation, and storage are important aspects for any serious grower.

The tasks associated with harvesting cannabis flowers vary by stage of processing but fall chiefly as the responsibilities of the farmer. Careful handling is imperative to preserve the integrity of the crop at each step, from cutting, to the trimming to curing before the flower buds are ready for storing, transportation, evaluation, packaging, and final sale.

The benefits to good harvest standards and storage methods are numerous. Flowers (buds) that are properly prepared look better, smoke better, and are easier to sell. Influenced by environmental factors, useful storage and packaging procedures help to protect the inherent properties that give a crop distinction and worth. Maintaining flower quality, THC, and CBD values along with terpene profiles rely on good harvest and preservation practices; vital to farming success.

9 | a. Preparing Plants Before Cutting
Flush for Content and Flavor

In the pursuit of a premium product, it is very important to take a few final steps before the actual harvest begins. By cleansing, or leaching the mineral impurities from the root zone, the flower formations begin to refine, and finesse to perfection when cutting time arrives, only days away from these important tasks.

A final step prior to removing flowers or stems from the plant will help to ensure that you maximize the quality and cleanliness of your harvest while bringing out the marvelous taste and aroma of your product.

There is an opportunity to enhance your premium cannabis simply and economically by flushing (also called leaching) the growing medium just before harvest. In a cultivation sense, it is the cleansing process just before harvest that improves the finished bud. It removes remnants of fertilizers (organic or inorganic) that have built up in the substrate or plant tissues that lead to a nasty taste in the crop. Not much attention goes to flushing and it is an easy step to miss, but it underscores the importance of specific plant care at every stage of development for optimal results.

Near the time of harvest, trichomes (resin glands that contain THC) are ripening and after much labor and expense, every grower eagerly waits for cut and trim time. By closely watching the flower development, you can expertly stop all fertilization and begin to provide the plants with straight, unadulterated water (distilled or reverse osmosis is ideal) or water containing a dilution of a commercial flushing formula usually containing citric acid. The easy procedure makes a remarkable difference not only in appearance of your premium flowers, but most importantly, the smell (nose) and flavor.

FLUSHING

The last five to ten (5-10) days of blooming before harvest is an excellent period to "clarify" cellular composition in your cannabis plants by removing salts and other contaminants from the growing medium and thereby from plant uptake, although some farmers begin fourteen (14) days before harvest. Salts are harmful to plant cells and high concentrations in the growing medium typically occur at the end of the plant life cycle following repeated fertilization during vegetative growth and blooming, especially without regular flushing. The most difficult part is determining when that window of opportunity begins. It can vary by plant within the same grow project and can change quickly based on light, climate, nutrients, and irrigation.

With so many variables including strain types, the best method of determining when to stop adding bloom formula fertilizers and additives such as molasses, etc., is to watch your plants closely starting at week six (6) or seven (7) of the bloom cycle. When the leaves, stems and flowers look loaded with trichomes, while most are light green and just before they begin to turn milky or translucent in appearance is usually a good time to start the flushing process. When most buds are very frosty in appearance, you are very close to harvest and flushing now, a few days before picking may be less than perfect, but the effort may still be worth it. Be aware that once cut, a stem or branch is no longer alive; all growth, and development in the cut portion stops.

Hydroponic growers or container growers will use periodic flushing as a method of removing unwanted nutrient salts that tends to build up in growing mediums throughout the growing cycle; a common cause of crop failure. Salt build up can be apparent with a white residue on the surface of the growing medium, but do not rely on visibility as an accurate measurement. Excessive salts shrink, damage, and destroy plant cells that exhibit first as wilting. This is harmful to the entire plant and irreversible. Some manufacturers of flushing formulas have suggested intervals for this type of flushing, but it usually occurs about every ten to fourteen (10 - 14) days in hydroponic culture.

Once final flushing has started prior to harvest, do not add any more fertilizers, additives, or chemicals to the planting medium. The last stages of the plant's life start with flushing near harvest time; do not interrupt the process once it begins. It is also prudent to reduce the humidity now within the growing environment if cultivating indoors or in a greenhouse and even more so at night. This protects the trichomes from collecting moisture droplets that can encourage devastating diseases that spread quickly in wet or over damp conditions, especially on trichome laden flower buds.

Note: Flushing does not remove bad pesticides not approved for food crops. Keep out the harmful chemicals for the entire life of your plants if you want a premium crop of value, integrity, and able to pass contaminant testing.

The simple step of washing out the obnoxious chemical tastes and odors prior to harvest results in a cleaner, better crop, worthy of all your hard work. Use a calendar, if needed, to chart when your crop should be ready for flushing and prepare ahead of time. When it comes time to fire up your herb, the extra effort right before harvest will be well worth it and the good reputation of a clean product from your farm, underscored.

Sometimes the most basic solutions are the best when working with cannabis and for drying, clothes lines and clothes pins are excellent examples. Cost-effective, usually portable, and easy to maintain, clothes line systems are a good way to hang cannabis branches for drying in most locations. Take advantage of floor to ceiling space with tiered lines and use the clothes pins for branches that have no easy hook method. Individual buds are best dried on mesh screens, but for a quick way to convert the use of a room, clothes lines work well for buds that remain on a branch or large stem.

Protect and Monitor as Buds Mature 9 | b.
Peak of Bloom Vulnerabilities

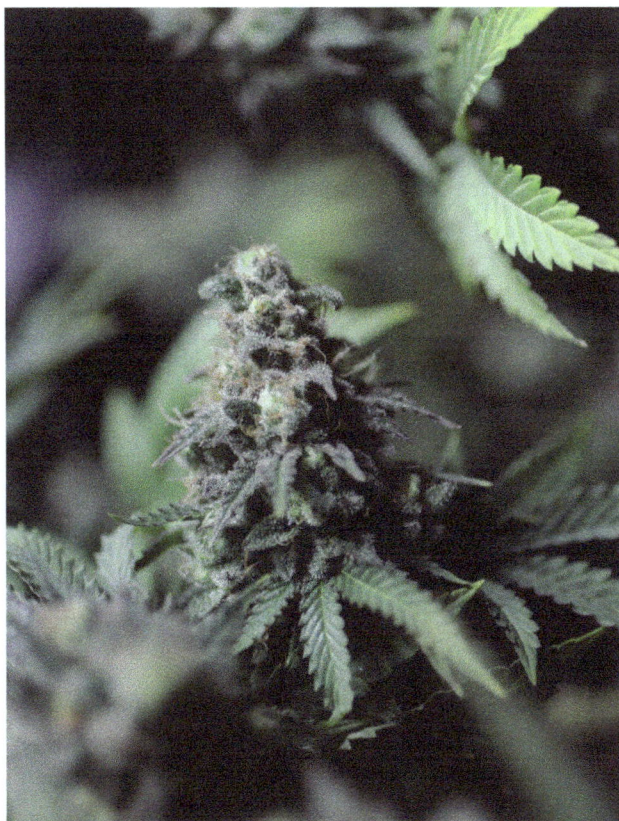

Analogies that use comparisons in nature are often best to describe farming concepts. So, when flowers ripen and crop maturation approaches, every effort must focus on protecting your cannabis plants just as a bee might protect her hive. Attention to every detail and nuance is a big part of harvesting success. Just like a bee, a watchful eye, care when needed and protection from pests and thieves are the primary responsibility for marijuana farmers as their frosty prizes become ready for cutting.

Experienced cannabis farmers will tell you that the definition of heartbreak is to lose your premium crop in the last days of maturation to avoidable damages. While they may appear strong and sturdy near harvest, marijuana plants are very vulnerable at the latter stages of development. Hybridization, culture techniques and sophisticated nutrient systems all contribute to robust plants that would not occur naturally and need attention to harvest for best crop production. Mature marijuana plants are also attractive to pests, diseases, and thieves. Fortunately, there are things you can do to protect your hard work and investment.

STRUCTURAL SUPPORT

The result of good cultivation techniques and growing standards will lead to heavy flowering and trichome development. This added weight requires supporting with some sort of staking or netting to avoid cracked stems and branches; rarely can you grow and harvest large bud formations, even in natural settings, without providing a reliable support system.

Plan and do the supporting work and installation before the plants need it. Waiting until cannabis plants are immense or with huge buds is often not in time since late staking usually results in damage or breaking of buds, stems and branches. Place netting early in the vegetative stage so plants can grow through it naturally. Trying to place netting over existing plants is challenging and will lead to plant damage; do not even attempt it when plants are mature.

CLIMATE CONTROL

It may not be possible to influence the climate to any degree if you are growing outdoors, or it may not be practical to control the environmental conditions to precise measurements if you are growing indoors, but the closer you can achieve "ideal" the better your crop will be. The biggest factor will be keeping rain, hail or fog off the maturing flowers if growing outside and reducing the humidity if growing indoors.

PEST AND DISEASE PREVENTION

Good cultural practices from the start are essential to a premium crop and particularly critical at the last stages of plant development. Healthy plants produce a better crop and make the plants more resistant to pests and diseases, but a host of factors that include the flower structure itself makes marijuana buds a target for attacks by pests and diseases that harm both the flower quality and yield or completely wipe out an entire garden.

Start by limiting the exposure of your plants to outside influences including people. Sure, you may be proud of your crop and wanted to show it to trusted souls, but unintentional contamination can take place from pests that hitch a ride on shoes and clothing. Take a photograph, instead.

Closely monitor the plants as flowers ripen for harvest and look for pests like russet mites, spider mites, broad mites, whiteflies, and other arthropods that love to infest marijuana flowers and plants at maturity. Follow strict guidelines for eradication and control and avoid harmful chemical solutions. Unless a pesticide is safe for use on food crops up to harvest, do not use it on your mature cannabis plants or you will ruin your product for consumption.

Although handling your plants should always be minimal, trimming, and pruning continues almost to the day of harvest. Avoid touching the trichomes or damage will be irreparable. Lastly, wear pH balanced nitrile gloves for any plant work.

Good air flow is imperative in the pest prevention process. Stale, damp air is an invitation to disease like *Botrytis* that spreads like wildfire through a marijuana garden. A constant air flow helps reduce this threat greatly within indoor growing spaces. Humidity ranges should be on the dry side and reduced to 20% at night close to harvest. Just as harmful in the last two (2) weeks of bloom is rain, hail, or fog along with the accompanying high humidity ranges when you grow outside. If water droplets collect around flowers and other plant parts, they can remain while harboring and spreading pathogens to your plants. Covering outside plants during wet weather periods is dangerous to your plants if not done correctly, so use great care to prevent water collecting on the cover and breaking the plants below, or from a falling cover on the plants in windy weather. Also make sure that covers do not drip condensed moisture onto your plants.

GUARD FROM THEFT

As many marijuana farmers have experienced, loose lips sink ships, or here, the wrong people who know about what you are growing put your cultivation project at risk. The best advice is to keep your farm or garden confidential from those that do not need to know. If permitting or inspections is a requirement by regulatory agencies, be professional and discrete about all your disclosures.

The security issues and best practices surrounding marijuana and cultivation of cannabis are beyond the scope of this book; however, it is a very smart practice to stay close to your crop when harvest is near. Round the clock monitoring or surveillance by cameras helps to discourage theft, but, personal safety is the most important thing so be smart and take no risks.

Close-up of a cannabis flower ready for harvest, showing trichomes and pistillate hairs.

Finishing Buds

You have reached a magic moment when the work is almost complete and harvested buds are ready for manicuring before drying and curing. Or, the buds are dry and ready for their final preparation before packaging for sale. Either case, the careful handling and meticulous trimming can make or break the commercial viability of your product. Trimmed properly, the flower buds have value exceeding that of under or over-trimmed buds and if damaged, will be good for extraction only.

Although various tasks in cultivation do not require huge degrees of skill, bud trimming does. Many successful cannabis growers who produce higher-end crops employ people who are very skilled and trim buds for a living. Your relatives may have good intentions, and friends may want to help, but lacking the skill set to trim your harvest quickly and efficiently, they can cause more damage than good.

The window of opportunity for trimming is very limited; do not underestimate the labor hours required to trim the buds soon after harvest. Unless trimmed immediately and properly, or hung for drying promptly, cut branches of mature flowers will limp and mold very quickly, making final preparation impossible and the value of the buds worthless.

Locating qualified labor varies by locale; but not difficult if you ask around. Once you find and lineup the labor pool, make sure they will be available when you need them and know up front what expectations are, especially for the rate of pay. Prepare to accommodate transportation if required, as workers may not have access. Make sure to have everything ready that they will need; trimming scissors, alcohol for cleaning the blades, trays for finishing and buckets for the leaf debris.

The next step is to keep the employees happy! From experience, believe that tunes, good food, and a comfortable work space with ergonomic seating make all the difference in the world for the quality of trimming and the finished product. Like all of us, bud trimmers appreciate respect that usually results in a greater effort for a current harvest and loyalty for the next time you need a good finisher for your worthy and beautiful crop.

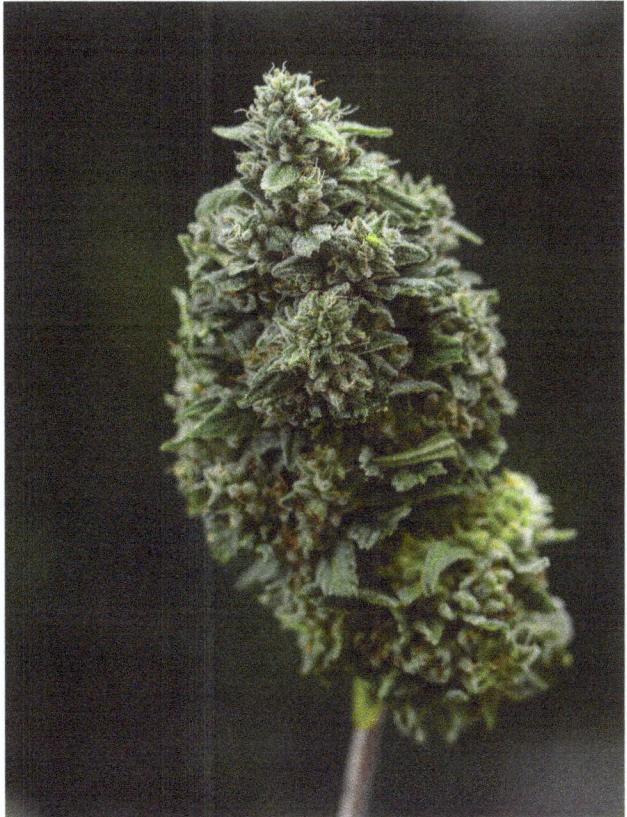

Even with mechanized trimmers available, cutting a premium crop is mainly hand labor that requires skill to ensure the quality of the finished product without damaging it in the process. The final steps are tedious and require proven and reliable protocols to harvest and prepare your premium quality marijuana flower buds, but the payoff is a stunning piece of artful work, fit for any lover of fine herb.

The careful labor of the premium weed farmer pays off when the time of harvest arrives. Swollen flower buds encrusted with white frost signal that yet another critical step in the cultivation process will soon begin. Like all parts of marijuana farming, everyone has personal preferences regarding the best methods of picking the ripened flowers. Despite the way you do it, a few important steps will help reduce the work involved and preserve the essential aspects of your crop; appearance, taste, aroma, THC, and CBD content, etc.

TIMING

The goal of any plant farmer is to harvest when the crop is at its peak. The same is true for cannabis growers who must carefully watch the maturation of the flower buds so removal from the plant occurs when most of trichomes are "ripe". The best way to describe this period is when most (80% or more) of the trichomes appear milky, instead of clear (immature) or golden brown (over ripe). There is a narrow window of prime time to pick your crop, about five to seven (5-7) days, so frequent examination of sample trichomes under magnification is essential.

MANAGING MOISTURE

A good way to help the harvested crop dry more efficiently is to withhold water from the plants a day or two before harvest if there is no wilting of leaves. Inside growers should drop temperatures below 80°F. and reduce the humidity to below 50%. Remove any large or yellowing leaves before harvest while the plant is in the final stages of bloom, a process that saves time later. Trichomes on this removed material are really beyond any value since the THC content has degraded and old leaves may bring pathogens into the grooming/manicuring room after stem cutting; discard this green waste according to local regulation. It is possible and practical to harvest only mature flowers and leave remaining buds to develop until ripe. Sometimes this is a day or up to a week depending on growing conditions and the plant strains. Make certain to have time and labor available for this technique.

MAKE THE CUT

Start the harvest at dawn or after a dark period. The next step is to remove the branches, colas, or stems from the plant. Accomplish this with sharp pruning shears, by cutting at the base of the stem, near the trunk. Removing individual flower buds in the field or the garden is not prudent since they will dehydrate quickly with damage likely from over handling. Remember that the trichomes are fragile and your goal to preserve their integrity through the drying process. Crushing them will create harm to the product by quick degradation and encourage irregular drying or mold.

TRIM EXPEDITIOUSLY

Make sure to have the workforce to trim your buds soon after removal from the plant and cut only what is practical to finish in a six (6) to twelve (12) hour period. Manual bud trimming takes about five (5) hours per pound while electric trimmers take about an hour to process per pound. Keep the harvested stems protected from the sun and temperature extremes and start trimming, or grooming of the flower buds promptly. After that, the excess foliage becomes limp and more difficult to remove quickly and efficiently. This type of work is best suited for indoors, at room temperature or cooler.

Equipment for trimming cannabis flowers does not need to be fancy. Roasting pans and a paper covered table makes it easy to manicure unwanted leaves before drying and curing.

There are farmers who prefer to harvest their crop, dry it, and then trim it before a final cure. This method has pros and cons; the chief drawback is that it may require more labor with varying results than the trimming when the unwanted foliage is still fresh, since damp foliage may encourage mold if left on the bud and becomes difficult to trim away when it dries and envelopes the flower. (Any moldy buds should be immediately discarded.) Most growers prefer to trim soon after the stem cutting and removal from the plant. Whatever method, trimming your harvested flower buds is a finishing step that you should become an expert in.

Professional bud groomers and trimmers deserve an award. Their skillful cuts on a marijuana flower produce the desirable nugget that is the culmination of hours of your tedious cultivation and probably buckets of money along the way. Like the effect a diamond cutter can have on a precious stone, trimming cannabis flowers are one of the most important tasks in crop production, so bud trimmers are no less important. Unfortunately, not every farmer has access to trimmers and the labor pool for such services can be seasonal or too expensive to a small grower. Note that a few trimmers are good at doing it the wrong way, so here are some proven methods that most use.

There are various tools that are available for trimming cannabis flowers. Some are expensive; others are essential. Basic equipment should include a tray, lined with paper to collect the trim material and the best shears/scissors that you can find. Your local hydroponic store probably stocks the models that bud trimmers prefer, so check them out. You want shears pointed, sharp and comfortable for extended use. Blade cleaners are very helpful as residue from broken trichomes will quickly collect on the blades. Lubricants like hemp oil will reduce the resin deposits on blades and make removal more efficient. Rubbing alcohol is an excellent cleaner although some prefer products specifically made for cannabis use.

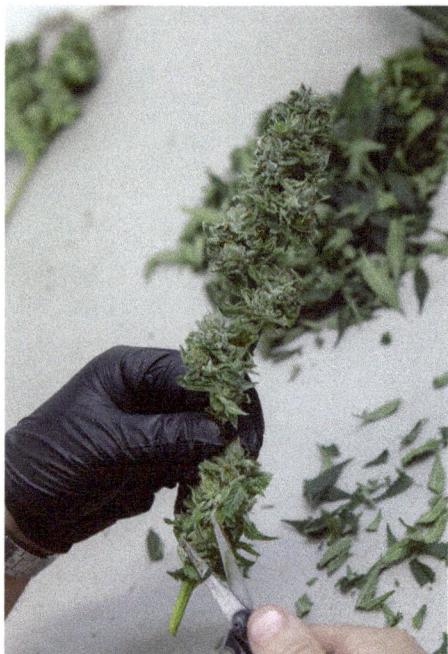

TRIMMING METHOD

Next, use gloves and handle the flower bud formations by their stalks or stems, avoiding contact with the actual flower structures. Rotate the stem and trim out leaf material, including the petiole or any excess foliage surrounding the bud. Regular grooming of plants during the vegetative stage or removing large leaves before harvest may reduce this task somewhat. This trimmed off material, when dried, has value for concentrates from cold water or CO_2 extraction. Set it aside for drying in paper bags.

When complete, place the trimmed flower buds in drying racks of breathable mesh, or like many farmers, if the buds remain on stalks, hang them on strings, or wires. Despite what myths you might have heard, the flower branches and stems hang upside down strictly for convenience and in protection of the trichomes. THC does not move into the bud region, or anywhere else for that matter. Consider that the least amount of handling during the entire finishing process is best for protecting the crop, despite the dry method.

Like the way most herbs cure, a slow process in a cool dark location is the best, with a gentle air movement to prevent mold from forming on surface condensation. The following section gives greater detail on drying.

Commercial cannabis cultivators may use electric trimming machines that are available to rent or own. They require some degree of skill to use properly but most offer instructional materials for use. Small, hand operated models are also available, but do not always provide consistent results. Any type of mechanical trimmer may cause damage to the buds, so testing and practice are highly recommendable. Also keep in mind that there is nothing comparable to a hand trimmed bud and many retailers and distributors will only purchase the hand trimmed flowers.

The final steps for the marijuana farmer who must quickly and efficiently harvest the prized crop can be some of the most important tasks when seeking quality results. Good organization of the work flow, the work space, supplies and the final destinations of the finishing crops are techniques to save both time and money, but more important are ways to protect the harvest and its value.

WASTE OR TRIM TO SAVE

A fair amount of material from the manicuring process has value, but some trimmed material has no use. To avoid sorting or worse, over handling the bud remaining on a stalk, a two-step process keeps what you can use or sell and discards the rest.

Removal of leaves and their attached stems are the first steps in a trimming process. Most, if not all the larger leaves from this step have little value, even with some trichomes present. Extractors do not want material that is not "clean" of debris like stems and big, dried leaves that lack enough resin glands to make cold water or CO_2 extraction feasible; this material is green waste. What they do want is the manicured material from the next step.

Procedurally, the trimmer clips the bud edges to leave a dense mass of trichomes on the remaining flower bud as depicted in the photograph below, typically before it dries. The material from this process has value and is suitable to dry for later use; after processing it for sale as "shake" or as bulk material for extraction to make a concentrated product. Small, fluffy buds, termed "popcorn" are also suitable for extraction and usually handled in the same manner as trimmed material from this step.

If there is no trimming before drying, some of the materials trimmed after bud drying may also be of value in extraction, although percentage compositions of trichomes are likely less than the preferred material described before. Usually, all the material trimmed off after bud curing in final manicuring before packaging or storing is suitable for combustible consumption or extraction, although in every trimming operation, there are variables that influence these general statements. The point is, most material has value, in varying degrees, while some material is not useable.

Consider the leaf material that you discard as green waste; it should go into local green debris recycling or composted with heat. It will develop mold quickly in bags or containers if moisture is present. Do not dump it or store it near your cultivation areas. Pests and diseases can survive for long periods on this material, making it unsafe to risk exposure to new or existing plants.

9 | d. Dry and Store Methods that Work
Five Steps to Perfection

Skilled herbalists make methodical effort and take great pride to refine their finished crop and so does the farmer of premium cannabis. The time and expense by the grower and magnificent effort from nature have produced the most amazing of all flowers, and it is up to the farmer to preserve the crop correctly and store it properly until use.

The same principles that herb and spice growers employ to preserve their precious commodities also apply for top-shelf marijuana farmers. With an array of equipment and supplies to properly preserve your harvest, good practices with excellent results are the expectation for cannabis farmers in any scale.

Step 1: Dry

The correct drying of marijuana should always be slow and methodical with slight ventilation. The objective is to let excess moisture gently evaporate from the plant material, much like drying other herbs. Oils and resins remain following the preservation process giving the finished product character and value.

There are two basic methods for drying marijuana, hanging from strings or wires, or placed in trays. The environmental conditions should be dark, within a certain temperature range on the cool side, while water moisture is exhausted or removed from the drying space with a gentle air flow.

The drying process may require several days, depending chiefly on plant material and the relative humidity. The point at which the stems snap, instead of bend is a good indicator that the buds are at a desirable level of dryness and ready for the next task in preservation.

Step 2: Cure

The best way to describe curing is to say it is a super-controlled period of refined drying in a stable environment. When bud trimming is complete and most are dry to a point where there is still some water moisture (never let them get totally dry or brittle) place them in airtight tubs, totes or containers in layers separated with a wicking-type of paper like paper-towel. Do not use newspaper; inks may transfer to the buds.

Your goal is to extract with precision the remaining unwanted water vapor, so open and "burp" the curing containers at least once a day until your buds are flawless and ready for safekeeping. The curing process may take a few days, but adds a great deal to the quality and shelf-life of your marijuana crop. Remember not to dry the material completely or it will ruin it and make it unsaleable except for concentrate extraction, at a low value.

Precise curing with two-way humidity packs available from hydroponic stores offers an effective way to maintain a relative humidity of 54% to 62%. With no special equipment or activation required for use, this simple technology originally developed for cigar humidors is an example of practical solutions for other products available for superior results in cannabis.

Step 3: Hand Manicure

If you plan to sell your finished flowers, make certain grooming is to perfection. A clean appearance of your manicured buds is important to marijuana buyers who scrutinize large amounts of product. There should be no crumbling and no excess material at the bottom of a storage jar, container, or bag for presentation. Loose or trimmed material, called "shake" has value for smoking or extraction; you just do not want it in large amounts when showing your flowers to a bud buyer.

Untrimmed buds require removal of leaf material after drying and before storage.

Professional cannabis buyers look at flower bud color, shape, structure, nose (aroma) and stickiness. Lastly make sure you are not selling giant stems, or "wood". Trim any excess dried stems and remove before packaging or storing. Cannabis sells by weight and the excess stems are unwanted "lumber."

Step 4: Test

Although most purchasers will test your crop on their own, typically a random amount from a one-pound sample, this is a useful step beforehand for the farmer when attempting to get your product reviewed or to evaluate the success of your crop. Make sure that you test with a certified lab in your area and follow their protocol for review of your crop.

Step 5: Store

Use the same standards used for spice and herb storage for your cannabis crop. Airtight containers placed in a cool, dark place is the rule. Options for practical and safe storage include turkey bags and airtight canisters, but the most popular is glass jars with airtight lids used in food preservation; available in various sizes, they are convenient, easy to clean and reusable.

No matter what expertise, there are times when the cure is inadequate with over wet material or too dry for intended use. Moisture packets that maintain 54%-62% humidity levels for stored cannabis flower buds are also useful for correction of incorrect levels. It is important to check your stored buds periodically, especially right after jarring, for any excess and unwanted moisture that may appear and lead to mold and rot if left unattended.

Literature on the subject suggests that the length of time that cannabis stores without substantial degradation can be a couple of years. The average consensus among dispensaries surveyed in California is that the ideal is within two months of harvest to maintain original texture, taste, aroma, and color, but outside ranges indicated best use is possible within six (6) to eight (8) months.

A careful and planned approach to preparing your crop for market underscores the pride of producing the finest result, every time you grow the most extraordinary member of the kingdom *Plantae*. With the fruits of your labor in safekeeping, it will be your reputation as a premium grower who gives it the long-lasting integrity it deserves.

IDEALS FOR DRYING CANNABIS FLOWER BUDS

Temperature	65° to 78° F (18.34° - 25.56° C)	Cooler is better
Humidity (rH)	40% to 65%	Dryer is better
Light	Dim to Dark	Darker is better

STORAGE STANDARDS FOR CURED CANNABIS FLOWER BUDS

Temperature	Less than 65° F and above 40° F
Humidity	54% to 62%
Light	Darkness

Off Season

Unless you grow indoors, 365 days a year, there are periods of downtime that you can use to your advantage. If you are an outdoor or seasonal grower, there may be weeks when you are not in cultivation mode. It is tempting to do as little as possible after all the work of a growing season, but after some rest, consider a couple of ways to build your business and have fun also.

When everything is clean, neat, and organized around your grow arena, it is time to go out and meet new people to help, and who can help you. Do not know where to start? Begin at every dispensary, plant nursery, or hydroponic store in your area. You do not need to buy anything (although you probably will); just go in and shop, introduce yourself and ask questions. The folks who work in this industry and serve growers are awesome people that you should know, and support.

Next on the to-do list in your off-time is to investigate and join professional organizations that will become an integral part of your farming network over time. These groups are very helpful as sources of information of great interest to growers; news, trends, policy decisions, and the lists go on. Guaranteed is the likelihood you will make lifelong friends when you enlist in an industry group.

A small, but significant action is to subscribe to as many cannabis related publications that you can afford. Here again, great people behind these efforts, working hard to support an emerging industry and very worthy of your patronage. The information is out there to help you, brought to you by people who really care and this is an excellent way to get it.

Shopping, joining, and reading. Sounds like the sorts of things any hardworking cannabis farmer deserves.

"Like the soil, mind is fertilized while it lies fallow, until a new burst of bloom ensues."
-John Dewey, Art as Experience

Chapter 10
Competing in the Green Rush

Farming is a business despite what you grow. Cannabis cultivation is no different and to be a successful grower, you need to be smart in everything you do, especially if you intend to be a producer of premium crops. As legalization increases, so does the intensity of the competition, so start right now to run your farm like a business.

As markets for legal marijuana open, for medical or recreational use, growers face the same challenges of production and pricing that are inherent in any crop commodity. Too much production and the price will fall; while too little production drives the price up temporarily, but with nothing to sell, there is no benefit to the farmer, retailer, or consumer in the long-term.

Alongside the market parameters are the all too common worry of small growers; how to compete with the big farms that get their pictures in the paper or their facilities featured in documentaries. After all, the images of mega-sized commercial projects can be very intimidating to a more modest farmer, or one that is just starting. Some potentially awesome growers never begin the journey with a belief that they cannot compete. Others loose interest and throw up their hands before giving a chance to realize a profit.

The grass is no greener for a big grower. Large cultivators face daunting bills in land, utility, and labor expenses that smaller growers do not. During the rush to riches in cannabis wherever it becomes legal, investors and operators of large cultivation sites also face stiff competition with one another or in a part of the market that only grows crops for extraction to make concentrate products.

No operation, despite size has all the advantages or disadvantages, so do not judge your farming opportunities based on plant numbers you have in production. Given the regulatory fees and taxes related to production that may be specific to your locale is a cost that all growers pay, your chief gauge of profitability is what each plant produces in weight and quality and what each plant cost to grow from clone to harvest. That basic formula applies to all growers, no matter how big or small. So, in principle, anyone can make money in legal cultivation; you just need a good plan, some wisdom about growing, the right mindset, and much dedication.

It really is a level playing field if you consider the basic goal of a success-driven cannabis cultivator; produce the best crop for the least amount of money and you are right where you need to be as a farmer. As a business person, you need to take it a few steps further though; scrutinize what it takes to produce your crop and be the best marketer the industry has seen. Sound impossible or difficult? Not at all, it is where the fun begins.

GET ORGANIZED

The best time to start organizing your operation is right now. It is a task that requires your attention, especially when you begin to bring others on board to provide professional help or labor duties for your business.

There is a ton of information available for small business startups that also applies to larger firms already in operation. Tap into those sources that might be able to help your farm and it does not need to be cannabis specific. Setting up or reorganizing is something that requires choosing a business type, getting it registered with the right authority, record keeping, and so on. If you are like many farmers, you probably would rather be digging in the soil or tending your plants, but paying attention to these details early-on makes it easier down the road when you become too busy with success to make it a priority.

GET ADVICE

Cannabis trade groups often host seminars or shows that are useful to any size grow-er. Industry specialists in all parts of the business, from regulators, to accountants, to marketers; they are all there eager to help you succeed in your enterprise. Many offer services to your specific locale and it is a great way to meet the people behind the scenes and geared to our vocations.

Local sourcing for assistance includes recommendations from others in the same line of work; another good reason to network.

BRANDING

This is a part of cannabis farming that you must think about and work on always. Your brand and the recognition from it are vital to the future of your success, despite how large you are now or how you see yourself down the road. The days of growing weed in the hills for a local customer are long gone. This may be disappointing to some, but it is the direction of this sector of agriculture, and we need to embrace it to do well.

There are things to be aware of in choosing your company name or brand and the specifics related to trademarks as you come up with a great logo. Unless you are su-per skilled at graphic arts, use your creative urge in the arena to guide a professional or team to help with this, and talk to a trademark attorney if you need to. You get one chance to make a great impression with it all for your brand, so do it correctly, right out of the box.

PROMOTE YOURSELF

The suggested methods of promotion like social media, advertising, etc., are a good idea, but for a cannabis farmer, it comes down to your reputation and the quality of product you are known for. It is one thing to grow premium cannabis, but if you are a jerk, really who cares? The best ways to promote yourself in the cannabis trade, is to be farmer of integrity and grow the best crop.

ALWAYS STRIVE TO PRODUCE THE BEST PRODUCT

This book is all about how to grow premium cannabis, in both lay and technical terms. Much of the information is what I as the author and a farmer, wish I knew when first cultivating the miracle plant. It is my sincerest desire that this compilation of can-nabis wisdom will help individual farmers and companies, despite production size, achieve greatness, so richly deserved. When growers do well, it starts a chain of suc-cess, where an entire industry benefits with accomplishment we have yet to imagine. If that is even partially influenced by this writing effort, then I will be a happy man.

With the suitable equipment, the proper facilities, and the right knowledge, you can grow a great crop. To grow a truly premium crop, one you can be proud of and put your brand on, season after season, you need to be insanely passionate about it and make it happen; good luck in all your growing endeavors!

Conversion Formulas for Growers

Temperature

˚C to ˚F: (˚C X 9/5) + 32

˚F to ˚C: (˚F -32) X 5/9

Length

Millimeters (mm) X .03937 = Inches (in.)

Inches (in.) X 25.40 = Millimeters (mm)

Meters (m) X 3.2809 = Feet (ft.)

Feet (ft.) X .3048 = Meters (m)

Volume

Dry

One gallon = 0.155 cubic feet (cu. ft.)

1 cubic foot = 6.43 US gallons, dry

Liquid

Liters (l) X .26418 = US Gallons (g.)

US Gallons (g.) X 3.7854 = Liters (l)

Milliliter (ml)/29.57 = U.S. liquid ounce (oz.)

US Liquid Ounce (oz.) = 29.57 Milliliters (ml)

US teaspoon = 4.93 milliliters

US tablespoon = 14.78 milliliters

1 cup = 236.58 milliliters

Glossary

Definitions of Terms Useful to Cannabis Farmers

Here, you will find the uncommon words and terms used in this book. Beyond these definitions, readers should further explore any words, concepts, and additional subjects of interest in cannabis cultivation and production.

Glossary

ABC soil: Soil having an A, B, and C horizon.

Acid: Media or solution with a pH less than 7.0.

Aeration: A cultivation method of adding air or oxygen to a root zone or a nutrient solution.

Aeration (soil): The exchange of air in soil with air from the atmosphere. Substantially aerated soil contains air like that in the atmosphere; poorly aerated soil is higher in carbon dioxide and lower in oxygen.

Aeroponics: A hydroponic method of growing plants in which the roots supplied with necessary nutrients in an aerosol mist.

Agronomy: Soil and plant sciences used in agriculture to manage land and crop production.

Alkaline: Media or solution that has a pH greater than 7.0; also known as basic.

Amendment: Altering a soil or other growing media to improve the structure, nutrient content, water holding capacity, cation exchange capacity, or air porosity.

Anemophilous: Pollinated by wind-blown pollen.

Annual: A plant that completes a life cycle in a year or less.

Available water capacity (available moisture capacity): The capacity of soils to hold water available for use by plants.

Bacteria: Single celled microorganisms that can be beneficial or harmful to plant growth.

Beneficial insects/microorganisms: Organisms that work with plants to defend against pathogens and pests.

Biodegradable: The ability to decompose or break down by natural bacterial or fungal action.

Breed: Sexual plant propagation.

Breathe (plant): roots draw in oxygen; stomata draw in carbon dioxide.

Bud: Undeveloped branchlet or stem. In cannabis, a group of flowering calyxes on a stem or branch. Bud and flower used interchangeably in marijuana vernacular.

Bud blight (bud rot): A pathological condition from non-specific plant diseases that attacks buds or new growth, causing withering and necrosis; often from a fungus genus, *Botryotinia*.

Buffer: A Chemical agent that reduces shock or fluctuations that may cause harm to plants.

Buffering capacity: The ability of a grow media or solution to resist change in pH values.

CBD: Cannabidiol, an active cannabinoid with medical benefits but does not make patients feel "stoned".

Calyx: The sepals of a flower; pod structure that protects the male or female reproductive organs.

Cambium (lateral meristem): Layer of actively dividing plant cells responsible for growth.

Cannabaceae: Scientific plant classification of the family of marijuana.

Cannabinoid: Diverse group of chemical compounds in cannabis flowers, leaves, and stems.

Cannabinoid profile: Test results yielding measured ratios of cannabinoids.

Capillary water: Water held as a film around soil particles and in tiny spaces between particles.

Carbon Dioxide: (CO_2) Odorless gas is an integral part of plant growth, with the ability to affect photosynthesis at specific light levels.

Carbohydrate: Compound of carbon, hydrogen, and oxygen, building blocks for plant structure (cellulose) and deliver energy for plant growth. Sugars and starches are types used to feed root zone microorganisms.

Cation: An ion carrying a positive charge of electricity; soil cations include calcium, potassium, magnesium, sodium, and hydrogen.

Caustic: Capability of destruction or damage by chemical activity.

Cell (plant): Basic structural unit of plant tissue; plant cells contain a nucleus, membrane, and chloroplasts.

Cellulose: A complex carbohydrate; stiffens plant tissue.

Centigrade (symbol C°): Temperature measurement scale; 100 degrees is the boiling point of water; 0 degrees is the freezing point of water.

CFM: Cubic feet per minute.

Chelate: A molecule that bonds with minerals resulting in easier and increased uptake into a plant; also prevents interactions with other minerals that could make the desired mineral unavailable.

Chlorine: Chemical used in water purification.

Chlorophyll: Green matter required for production of carbohydrates by photosynthesis.

Chloroplast: Containing chlorophyll.

Chlorosis: Yellowing of a plant resulting from lack of chlorophyll production during photosynthesis; causes are chiefly pH issues, nitrogen deficiency, or iron deficiency.

Clay (soil): Fine particles of organic matter and minerals, densely arranged. Clay soils are not suitable for crop cultivation, or for growing plants in containers.

Climate: The average weather conditions in a grow area, greenhouse, or indoors.

Clone: The technique of asexual propagation by cutting and rooting, or the individual rooted cutting.

Color spectrum: The band of colors emitted by a light source, measured in nm.

Color tracer: Added agent for tracing fertilizers in solutions.

Compaction: Soil condition where densely packed allowing minimal root penetration.

Companion planting: Planting certain plants to discourage pest insects and to attract beneficial insects.

Compost: Decayed or decaying organic matter from plant or animal sources; provides nutrients needed for plant growth.

Cortex: Outer layer of tissue below the epidermis on a root or stem.

Cotyledon: The first set of "leaves" of a dicot seedling. Serves as the nutrient source until the plant develops roots for uptake of minerals.

Cover crop: Annual plants grown to improve and protect the soil between periods of regular crop production.

Critical day length: Maximum day length that will induce flowering in cannabis.

Cross pollinate: Pollinating two plants with different ancestry.

Cultivar: A cultivated variety; a plant species that does not occur naturally.

Cure: Drying cannabis for consumption.

Cuticle: Thin layer of waxy-like cells, occurring on the surface of plant parts.

Cutting: A stem portion of the plant for generation of a new, identical plant (clone) by rooting.

Damping Off: Describes the die-back of plant cuttings and seedlings caused by fungi and other factors; principally from a plant pathogen, *Pythium*.

Dehydration: Water loss from the plant leaves.

Deplete: A removal of available nutrients making soil infertile.

Desiccate: To dry up as in insect desiccation by soaps.

Dioecious: A plant that has male and female organs on separate plants; one plant produces flowers (female) and one plant produces pollen (male).

Dose: Amount of chemical added to a specific amount of solution.

Drainage: Method to empty soil of excess water.

EC meter: An instrument that measures the electrical conductivity of a solution or growing media.

Elongation: The extension of a plant as it grows taller and wider, creating greater distance between the internodes of the plant.

Eluviation: Movement of material within the soil from one place or horizon to another.

Equinox: When the Sun crosses the equator; day and night are equally 12 hours long in both Spring and Fall.

Erosion: Losses of the land surface by geologic agents like water, wind, ice.

Fahrenheit (symbol °F.): Temperature measurement scale; 212 degrees is the boiling point of water; 32 degrees is the freezing point of water.

Fallow: Agricultural land left idle to restore productivity through accumulation of moisture and decomposition of organic matter.

Fan leaf: Large marijuana leaf lacking potency.

Feed: Horticulturally, to supply nutrients to plants in a solution or in dry form.

Female (plant): Pistillate; ovule; fertile seed producing if pollinated.

Fertility: The quality that enables a soil to provide nutrients, in adequate amounts, and in proper balance, for the growth of plants.

Fertilizer burn: The result of too much fertilizer, starting with brown leaf tips and proceeding to curl and die-back.

Flat: A shallow tray used to start cuttings or seedlings.

Foliage: The leaves of a plant.

Foliar Feeding: Spraying or misting plant leaves with readily absorbable nutrients.

Fulvic acid: Liquid portion of humus that can increase nutrient uptake by plants.

Fungistat (fungicide): Product to kill or inhibit fungus.

Fungus: A plant form lacking chlorophyll; mold, rust, mildew.

Gene: Inherited through propagation, part of a plant chromosome that determines potential production and resistance to disease.

Genotype: Specific genetic composite of an individual plant.

Girdling: Damage to a stem or branch causing a restriction in water and nutrient flow.

GPM: Gallons per minute.

Ground water: Water filling all the unblocked spaces of material below the water table.

Gynoecium (pistil): Female part of a flower.

Gypsum: A naturally occurring mineral used to lower soil or growing media pH.

Hardening off: When a plant adapts to a natural, outdoor environment from a controlled, indoor environment.

Hemp: A tall growing variety of cannabis cultivated for industrial uses.

Hermaphrodite: A single plant having both male and female organs.

HID: High intensity discharge bulbs/lights.

Horizon: Soil layer having distinct characteristics produced by soil-forming processes.

Hormone: In a botanical context, a chemical that controls the growth and development of plants. Root inducing hormones helps cuttings (clones) root.

Horticulture: A branch of agriculture concerned with the art, science, technology, and business of growing plants.

HPS: High pressure sodium bulbs/lights.

Humidity (relative): A ratio of the water in each volume of air at a given temperature and the greatest amount of water the air could hold at the same temperature.

Humic acid: The liquid portion of humus that comes from decomposed organic matter.

Humus: The decomposed, stable part of the organic matter from plants and animals in soils.

Humidity: Ratio between the moisture in the air and the greatest amount of moisture the air could hold at the same temperature.

Hybrid: The resulting plant from two plants of different strains or varieties.

Hybrid vigor: The increased health, strength, and growth rate offspring plants over that of the parents.

Hydrogen: Colorless, odorless gas that combines with oxygen to form water.

Hydroponics: Cultivation of plants in nutrient solutions without soil.

Hygrometer: An instrument for measuring relative humidity.

Inbred: Plants from the same breed, parent, or ancestor, typically grown from the same seed lot.

Inductive photo period: The length of light required to begin flowering.

Inert: A solution or growing media that is chemically non-reactive.

Irrigation: Application of water to soil or growing media in crop production.

Insecticide: A Compound that kills insects.

Kilowatt hour: The Measure of electricity used per hour (1,000 watts).

Land race: A naturally growing (wild) strain of cannabis.

Leaching: In crop production, dissolving, or washing out soluble salts from a soil or growing media.

Leaf curl: Leaf edges curl under; typically caused by incorrect watering, fertilization or the result of insects or disease.

Leaflet: Small, immature leaves.

Leaves: An External plant part attached to stems and branches, absorb light energy for use.

Leggy: Sparse foliage with abnormal internode spacing. Often caused by lack of blue light, CO2, or excessive nitrogen.

Lime: A mined mineral compound that will raise the pH of soil or growing media.

Litmus paper: A chemically sensitive paper used to measure pH when compared to a color scale.

Loam: A type of soil consisting of clay, silt, sand, and organic matter.

Lumen: A unit of measurement of light; used to rate the output of light based on the input of electrical watts. One lumen equals candlelight intensity one square foot away from the source.

Macro nutrient: Major nutrients required for plant growth: nitrogen, phosphorus, potassium, calcium, magnesium.

Manicure (pruning): Removal of cannabis plant portions to encourage growth of THC potent sections.

Meristem: The region of a plant where active cell division and new growth occurs.

MH: Metal Halide.

Micro nutrient: Minor nutrients required for optimal plant growth; iron, boron, copper, molybdenum, zinc, manganese, nickel, and cobalt.

Moisture meter: An instrument used to measure the moisture in soil or growing media.

Mother plant: A cultivated plant of desirable characteristics, used for producing cuttings (clones) for successive crop production.

Mulch: A layer of organic material used as a protective layer to retain moisture and inhibit weed growth outside; inside use is problematic and may encourage harmful fungi.

Mutation: A heritable and sudden change in the genetic material.

Nanometer (nm): 0.00000001 of a meter; used as a scale to measure wavelengths of light spectrums as used in describing plant growing lights.

Necrosis: Localized death of part of a plant.

Node: The joints on a plant where leaves attach.

Nutrient: A plant food; a mineral that a plant utilizes to survive and grow; from the soil (media) through roots and from the atmosphere through leaves.

Open pollination: Natural pollination without human controls.

Organic: Often used as a marketing term; in agriculture, it refers to natural process and materials used in crop cultivation and without synthetic substances. In chemistry, it refers to a substance containing carbon.

Organic matter: Plant and animal material in the soil in various stages of decomposition.

Ovule: The "egg" of a plant found in the calyx containing female genes. Fertilization of the ovule will grow it into a seed.

Oxygen: A colorless, odorless element required in soil to sustain plant life.

Parasite: An organism hosted by another organism. Examples include mites, fungus, etc.

Passive: A type of hydroponic system that transports nutrients to the plants through absorption.

Pathogen: A disease causing organism; bacteria, virus, fungi, etc.

Peat moss: Sphagnum is a genus of 120 species of mosses; partially decomposed moss vegetation from habitats in cold and moist regions harvested for agricultural use. Peat may hold 15-25 times its dry weight in water. Useful as a soil conditioner.

Perennial: A plant that completes a life cycle over several years, like a long-lived herb or shrub.

Pericycle: Regulates the formation of lateral roots.

pH: A measurement of the acidity or basicity of a solution; the power of hydrogen. On a scale of one to fourteen (1-14); free hydrogen ions in a solution. The optimal range for plant growth is 5.5 to 6.8.

pH meter: An instrument that measures pH values.

Phenotype: A visible feature of an individual plant that results from an interaction between the genotype and the environment.

Phloem: Plant tissue that transports food and water.

Photo period: The length of light and dark expressed in hours during a 24-hour period.

Photosynthesis: The conversion by plants of light (energy) to chemical (energy). Energy from the sun or an artificial light source, water, and carbon dioxide build carbohydrate molecules such as sugars.

Phototropism: A specific movement of a plant to light.

Phyto-: Denoting plants.

Pigment: Light absorbing substance.

Pistils: A part of the gynoecium with an ovary, style, and stigma where pollen attaches. In cannabis, a pair of white hairs from the top of a female calyx.

Pod: A dried calyx containing a seed.

Pollen: Fine dust-like spores containing male genes.

Pollen sack: Portion of male flower contains pollen.

Pollination: Fertilization of a seed plant by pollen.

Primary nutrients: Expressed as N-P-K; nitrogen, phosphorous, potassium.

Productivity (soil): The capability of a soil or grow medium for producing a plant or sequence of plants by specific crop management.

Profile: A vertical cross section of the soil extending through all its horizons.

Propagate: To germinate a plant from seed (sexual propagation) or grow a new plant by taking a cutting from an existing plant and rooting the cutting (asexual propagation).

Prune: Changing a plant shape or growth structure by cutting leaves, stems, shoots, flowers, or fruits to control and direct growth.

Recovery: A type of hydroponic system that retains the nutrient solution for reuse.

rH: Abbreviation for relative humidity.

Rhizosphere: The area around the roots of a plant where beneficial microorganism lives and the region where nutrient processing and disease suppression occur.

Rock wool: Also known as mineral wool, a manufactured growing media from fibers created by heating volcanic rock at high temperatures. A naturally high pH requires conditioning before plant use. (Also spelled rockwool.)

Root bound: Roots of a plant restricted from normal growth, usually from the confines of a container or from a soil layer that greatly restricts roots.

Root zone: The region of soil or growing medium penetrable by plant roots, or where they exist.

Runoff: Water from precipitation or irrigation that flows off the surface of the land without penetrating the soil (surface runoff) or water that enters the soil before reaching surface streams (ground-water runoff).

Salt: Crystalline compound whose cation comes from a base whose anion comes from an acid. Buildup of salts in soil or growing media can cause harm to plants by burning them and preventing nutrient absorption.

Sea of Green: A cannabis cultivation method where tightly spaced clones flower immediately for fast production with no wasted growing space.

SCROG: Screen of green; a cannabis cultivation method using netting, trellising, etc.

Secondary nutrients: Calcium (Ca) and magnesium (Mg).

Series (soil): A group of soils with similar profiles (composition, thickness, arrangement) except for differences in texture of the surface layer or of the underlying material.

Sinsemilla: In Spanish, without seed; used to describe female cannabis plants grown without exposure to pollen and lacking subsequent seeds.

Soilless mix: Used to describe a growing medium often containing organic material like moss and mineral products like vermiculite, perlite, or pumice.

Soluble: Dissolvable in water.

Solution: A mixture of solids, liquids, or gases.

Spore: Offspring of a fungus, algae; seed-like.

Stamen: Male flower organ; pollen producing.

Starch: A complex carbohydrate; manufactured by plants and stored in plant tissues.

Stipule: Small appendages on the base of leaf stalks.

Stomata: A small portal in the underside of leaves of a plant where nutrients absorb, water releases, and CO_2 consumed.

Strain: A lineage of cannabis with the same characteristics.

Stress: Chemical or physical factor causing harm to plant requiring extra exertion. Insufficient water or high temperatures are examples of stress causing factors.

Substrate: A growing medium for living organisms.

Sugar: A simple carbohydrate produced by plants in photosynthesis.

Tap Root: The main root of a plant that grows from the seed. Lateral and branch roots grow from the tap root.

Taxonomy: Scientific classification of plants and animals.

Tepid (lukewarm): Warm water temperature optimal for chemical processes in most plant growth; 70 to 80 degrees F (21 to 27 degrees C). In cannabis, optimal is 68 degrees F.

Terminal bud: The bud growing at the tip of a stem.

Terpene: Hydrocarbon found in plant resins, with distinct and diverse aromas.

TDS: Total dissolved solids.

THC: Tetrahydrocannabinol; primary intoxicant (physiological) in cannabis.

THCV: Tetrahydrocannibivarol; psychoactive compound in cannabis.

Thin: To cull or remove plants to create more space for other plants; creates better air circulation and light penetration.

Transpire: The release of water and other vapors by a plant through stomata.

Trichome: Epidermal outgrowth; on cannabis, a resin-filled plant hair.

Ultraviolet (UV): Short wavelengths of light that exist outside the visible spectrum beyond blue and violet.

Variety: A specific and distinct phenotype or strain.

Vascular: A description of a plant circulatory system of water and nutrient transport.

Vector: An agent (insect, vermin) that transmits disease.

Vegetative: The growth stage in cannabis of new leaves and rapid chlorophyll production.

Wetting agent: A compound that reduces the surface tension of water; used in horticulture to increase water absorption by soil or growing media.

Wick: Part of a passive hydroponic system where the growing media and roots absorb water and nutrients in a slow process.

Wilting point: (or permanent wilting point). The moisture content of soil or growing media at which a plant wilts and does not recover.

Xylem: Plant tissue that moves water and nutrients from the roots to the branches, stems and leaves.

Bibliography

The increasing body of knowledge pertaining to marijuana should interest anyone who wants to grow the best possible plants with the highest yield. This book discusses key areas of cultivation to grow premium cannabis crops, but there are other aspects for serious growers looking for the best results that require in-depth knowledge. If we continue to research and learn more about cannabis and how to grow it better, great inspiration and discovery are sure to follow, not to mention better and better crops.

Cervantes, Jorge. *Marijuana Horticulture: The Indoor/outdoor Medical Grower's Bible.* Sacramento, CA: Van Patten Pub., 2006. Print

-

Cervantes, Jorge. *The Cannabis Encyclopedia: The Definitive Guide to Cultivation & Consumption of Medical Marijuana.* Van Patten Publishing: n.p., 2015. Print.

-

Clarke, Robert Connell. *Marijuana Botany: An Advanced Study, the Propagation and Breeding of Distinctive Cannabis.* Berkeley, CA: Ronin Pub., 1981. Print.

-

Davis, B.N.K. *The Soil.* London: Harper Collins, 1992. Print.

-

Green, Greg. *The Cannabis Breeder's Bible: The Definitive Guide to Marijuana Genetics, Cannabis Botany, and Creating Strains for the Seed Market.* San Francisco, CA: Green Candy, 2005. Print.

-

Green, Greg. *The Cannabis Grow Bible: The Definitive Guide to Growing Marijuana for Recreational and Medical Use.* San Francisco, CA: Green Candy Pr., 2010. Print

-

Jones, J. Benton. *Plant Nutrition and Soil Fertility Manual.* Boca Raton: CRC, 2012. Print.

-

Ohlson, Kristin. *The Soil Will Save Us: How Scientists, Farmers, and Foodies Are Healing the Soil to Save the Planet.* Rodale Press: n.p. 2014. Print.

-

Plaster, Edward J., and Edward J. Plaster. *Soil Science & Management.* Albany: Delmar, 1997. Print.

-

Rosenthal, Ed, and Kathy Imbriani. *Marijuana Pest and Disease Control.* Oakland, CA: Quick American, 2012. Print

-

Rosenthal, Ed. *Ed Rosenthal's Marijuana Grower's Handbook: Your Complete Guide for Medical & Personal Marijuana Cultivation.* Quick American.: 2010. Print

-

Thomas, Mel. *Cannabis Cultivation: A Complete Grower's Guide.* San Francisco, CA: Green Candy, 2012. Print.

Industry Associations and Organizations
(Check for local and State groups near you)

The American Herbal Products Association (AHPA)

8630 Fenton St. #918

Silver Spring, MD 20910

www.ahpa.org

Americans for Safe Access (ASA)

1624 U Street NW

Suite 200

Washington, D.C. 20009

www.safeaccessnow.org

American Trade Association for Cannabis & Hemp (ATACH)

www.atach.org

Cannabis Business Executive

CBE Press LLC

2750 Gallows Rd., Suite 344

Vienna, VA 22180

www.cannabisbusinessexecutive.com

Cannabis Business Summit & Expo

1501 India St. Suite 103-60

San Diego, CA 92101

www.cannabisbusinesssummit.com

Hemp Industries Association (HIA)

www.thehia.org

Marijuana Policy Project

P.O. Box 77492

Capitol Hill, Washington, D.C. 20013

www.mpp.org

Marijuana Industry Group (MIG)

1700 Lincoln Ave. Suite 1530

Denver, CO 80203

www.marijuanaindustrygroup.org

National Association of Cannabis Businesses

www.nacb.com

The National Cannabis Industry Association

126 C Street N.W. 3rd Floor

Washington, D.C. 20001

www.thecannabisindustry.org

National Hemp Association

100 M Street, SE Suite 500

Washington, D.C. 20003

www.nationalhempassociation.org

National Organization for the Reform of Marijuana Laws

1100 H Street, N.W. Suite 830

Washington, D.C. 20005

www.norml.org

Organic Cannabis Association (OCA)

www.organicca.org

Project CBD

www.projectcbd.org

The Global Store

There are horticultural suppliers that offer products useful to any grower of any crop. As new ideas, techniques, and products evolve in agriculture, the opportunities for application in cannabis cultivation develop as well, with a marketplace rich in selection. For the cannabis grower who wants a promising future, it boils down to obtaining the best value for every dollar spent and getting exactly what you want and need.

At the start, realize that the least expensive price on an item is not a measure of value. You must factor in all the benefits before assessing an expenditure and determine if a product or service is useful in your crop production and if a vendor selling it deserves your hard-earned money. It is also important to understand the advantage of working with companies and individuals regularly to establish business relationships that help on your road to achievement. Shopping strictly on price may feel good today but will not serve your interests, long-term without a support network and that includes all your regular suppliers and service providers.

It is always a good idea to shop around and constantly experiment with new products and ideas; it is the essence of competition in the market and it fuels innovation and new technologies as we get better at farming this incredible plant. The point not to miss is that in the shopping process, keep your eye on establishing a loyal relationship with those you like and trust.

Professional growers also understand that there are not huge markups on the merchandise that you will acquire. Manufacturers, distributors, and retailers work hard too, and the margins are always shrinking as more enter this booming marketplace. So, they count on regular customers to stay in business, which is why they make excellent "partners". Communicating effectively and remaining true to your vendors will serve everyone well.

Starting local and moving globally is a sound strategy in any procurement for your farm or garden. That means, support the brick and mortar stores in your geographic area for what they do best and then find sourcing on the Internet for what those companies do well. It is not an all or nothing kind of approach; every business and manufacturer serving this industry deserves a chance, so search until you find what is a good fit for your operation today and your aspirations for tomorrow.

Resources
Merchants and Manufacturers

For any book to be a functional reference tool, it should intrigue, inspire, and offer other avenues of information to explore. This directory incorporates those objectives with some of the many great companies who provide supplies, equipment and technologies that are useful in cultivating premium cannabis; all are worthy of your consideration.

(Brand names and trademarks belong to their respective owner(s). Web sites, addresses, and contact information may change after publication.)

Resources

• Hydroponic Distributors

BWGS Full Spectrum Distribution
   4045 Perimeter W. Dr. Ste. 400
   Charlotte, NC 28214
   www.bwgs.com

Humboldt Wholesale
   888-499-4353
   www.humboldtwholesale.com

Hydrofarm
   Petaluma, CA 94954
   800-634-9990
   www.hydrofarm.com

Sunlight Supply
   3204 NW 38th Circle
   Vancouver, WA 98660
   www.sunlightsupply.com

• Online Hydroponic Dealers

Hydrobuilder, Inc.
   312 Otterson Dr. Suite D
   Chico, CA 95928
   888-815-9763
   www.hydrobuilder.com

Planet Natural
   1251 N. Rouse Ave.
   Bozeman, MT 59715
   800-289-6656
   www.planetnatural.com

1000 Bulbs.com
  2140 Merrit Dr.
  Garland, TX 75041
  800-624-4488
  www.1000bulbs.com

• Fertilizers, Substrates & Amendments

Advanced Nutrients
  1625 Heritage St.
  Woodland, WA 98674
  www.advancednutrients.com

Aurora Innovations
  PO Box 22041
  Eugene, OR 97402
  866-376-8578
  www.aurorainnovations.org

Botanicare
  6858 W. Chicago Street Suite 3
  Chandler, AZ 85226
  877-753-0404
  www.botanicare.com

Canna
  11400 West Olympic Blvd. Suite 200
  Los Angeles, CA 90064
  www.cannagardening.com

Espoma
  6 Espoma Rd.
  Millville, NJ 08332
  www.espoma.com

FoxFarm Soil & Fertilizer Company
P.O. Box 787
Arcata, CA 95518
800-436-9327
www.foxfarmfertilizer.com

General Hydroponics
2877 Giffen Ave.
Santa Rosa, CA 95407
www.generalhydroponics.com

Growstone
Albuquerque, NM
505-908-9701
www.growstone.com

Heavy Harvest
113 Cherry St. Suite 68221
Seattle, WA 98104
970-444-2852
www.heavyharvest.com

Humboldt Nutrients
6 Fifth St.
Eureka, CA 95501
888-420-7770
www.humboldtnutrients.com

Maxsea
PO Box 640
Garberville, CA 95542
www.maxsea-plant-food.com

Plant Revolution Inc.
412 Goetz Ave.
Santa Ana, CA 92707
www.plant-success.com

Premier Tech Horticulture
    200 Kelly Rd. Unit E-1
    Quakertown, PA 18951
    800-525-2553
    www.pthorticulture.com

Royal Gold Potting Soil
    1689 Glendale Dr.
    Arcata, CA 95521
    707-822-4653
    www.royalgoldcoco.com

Sunshine Advanced
    Sun Gro Horticulture
    770 Silver St.
    Agawam, MA 01001-2907
    1-800-732-8667
    www.sunshineadvanced.com

Vermicrop Organics
    5050 Arboga Rd.
    Olivehurst, CA 95961
    800-994-8775
    www.vermicrop.com

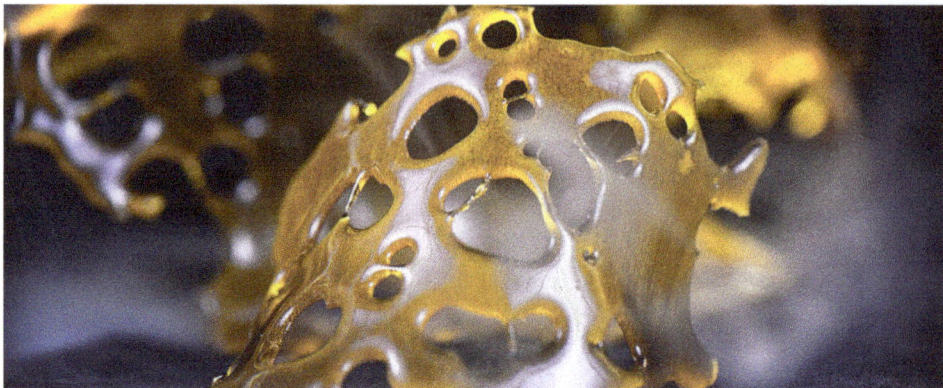

Your Niche

Most cannabis growers love what they do. For those that like it and are also good at cultivating premium crops, it is a way of life that words cannot adequately describe. Even so, there are other avenues aside from farming that are open to cannabis growers looking for the perfect niche.

Finding the right fit within an industry that is changing rapidly can be daunting with so many awesome opportunities; not only in cultivation, but in all the support activities as well. If you currently grow cannabis or plan to, use it as a learning opportunity and the chance to build a network of people who can help on your career path no matter where you end up.

There is no doubt that growing premium cannabis is a good way to enter the industry, but "You don't know what you don't know." Unless exposed to the many and varied jobs in the cannabis field, it is difficult to know what you might like along with or instead of, growing. There is distribution, sales, packaging, marketing, and the list goes on. So, immersing yourself in cultivating marijuana and producing a premium crop is excellent credentials for the experience needed for those yet, undiscovered dream jobs that are out there for you to explore. The industry will benefit from your knowledge; capitalize on that experience and get active within the organizations that support your livelihood. Often, it the set of keys that open other doors in the business of legal marijuana.

The reason to participate is simple; the folks who are looking for talent attend the events and meetings where seasoned growers belong. Recruiting from an organized group of like-minded individuals is smart and where many great people have found good positions. You may think, I am a grower, forever. Great, but never say never because you may develop or invent something that can change everything, or you may decide to work for a bigger organization, or teach, or who knows what, so remain open to the possibilities, always.

Finally, no matter what you do in, or for the world of cannabis, be the best. It is an enormous field offering truly rewarding work that can help so many. Exceptional achievement will always provide a significant contribution to a great cannabis occupation and a wonderful life, working with incredible people and the unsurpassed plants of our attention.

Accuracy, Completeness, and Timeliness of Information

This book made every effort to provide pertinent information for unspecific informational use, only. We; the author, publisher, assigned representatives or agents thereof are not responsible if information made available in this book is not accurate, complete, or current. The material in this book is for general information only and should not be relied upon or used as the sole basis for making decisions without consulting primary, more accurate, more complete, or more timely sources of information. Any reliance on the material in this book is at your own risk.

Internet websites listed in this work might have changed or disappeared after publication and company addresses, or other contact information might have changed. Any potential source of information listed for reader consideration does not mean that the author or publisher endorses the information, the company, or the website, or the products and information they may provide or recommend. Products listed for comparison value do not mean that the author or publisher endorses the product that might have changed and offers no guarantee of price or availability. Brand Names, Trade Names, Trademarks, and Service Marks, belong to their respective owners.

This book does not encourage the use of marijuana for any purpose other than to treat medical conditions allowed by law. The content does not represent the safety or efficacy of medical or recreational marijuana use, but rather promotes high standards of farming for clean, safe crops and environmental protection where legally grown.

www.ingramcontent.com/pod-product-compliance
Lightning Source LLC
Chambersburg PA
CBHW042310210326
41598CB00041B/7340